Solid-Liquid Filtration
A users' guide to minimizing costs and environmental impact; maximizing quality and productivity

Solid-Liquid Filtration
A users' guide to minimizing costs and environmental impact; maximizing quality and productivity

Trevor Sparks

AMSTERDAM • BOSTON • HEIDELBERG • LONDON
NEW YORK • OXFORD • PARIS • SAN DIEGO
SAN FRANCISCO • SINGAPORE • SYDNEY • TOKYO

Butterworth-Heinemann is an imprint of Elsevier

Butterworth-Heinemann is an imprint of Elsevier
The Boulevard, Langford Lane, Kidlington, Oxford OX5 1GB, UK
225 Wyman Street, Waltham, MA 02451, USA

First edition 2012

Copyright © 2012 Elsevier Ltd. All rights reserved.

No part of this publication may be reproduced, stored in a retrieval system or transmitted in any form or by any means electronic, mechanical, photocopying, recording or otherwise without the prior written permission of the publisher

Permissions may be sought directly from Elsevier's Science & Technology Rights Department in Oxford, UK: phone (+44) (0) 1865 843830; fax (+44) (0) 1865 853333; email: permissions@elsevier.com. Alternatively you can submit your request online by visiting the Elsevier web site at http://elsevier.com/locate/permissions, and selecting *Obtaining permission to use Elsevier material*

Notice
No responsibility is assumed by the publisher for any injury and/or damage to persons or property as a matter of products liability, negligence or otherwise, or from any use or operation of any methods, products, instructions or ideas contained in the material herein. Because of rapid advances in the medical sciences, in particular, independent verification of diagnoses and drug dosages should be made

British Library Cataloguing in Publication Data
A catalogue record for this book is available from the British Library

Library of Congress Cataloging-in-Publication Data
A catalog record for this book is availabe from the Library of Congress

ISBN: 978-0-08-097114-8

For information on all Butterworth-Heinemann publications
visit our web site at books.elsevier.com

Working together to grow
libraries in developing countries

www.elsevier.com | www.bookaid.org | www.sabre.org

ELSEVIER   BOOK AID International   Sabre Foundation

Transferred to Digital Printing in 2012

Contents

Preface — xi
List of Figures and Tables — xiii

1. **Introduction**
   1.1 Scope — 2
   1.2 Summary — 9

Part I
Background

2. **History of Filtration**
   2.1 Origins — 13
   2.2 Industrial Revolution and Onwards — 14
   2.3 Recent History — 16
   2.4 Current Trends — 17
       2.4.1 The Shape of the Industry Today — 18
   2.5 Summary — 19

3. **Physical Phenomena**
   3.1 Basic Narrative of a Filtration Process — 23
   3.2 Step-by-step Narrative — 25
       3.2.1 The Nature of a Solid–Liquid Suspension — 25
       3.2.2 Particles and the Filter Medium, the Very Early Stages of Cake Formation — 26
       3.2.3 Cake Growth — 29
       3.2.4 Cake Pressing — 33
       3.2.5 Cake Washing — 33
       3.2.6 Cake Draining — 38
       3.2.7 Air Drying — 39
   3.3 Other Notable Phenomena and Things to Look out for — 41
       3.3.1 Variations in Cake Properties — 41
       3.3.2 Rapidly Settling Slurries — 42
       3.3.3 Compressible Filter Cakes — 43
       3.3.4 Migration of Fines — 43
       3.3.5 Precoat and Body Feed — 43
   3.4 Summary — 44

Part II
Competitive Advantage

4. Competitiveness in Processing
- 4.1 Production Cost — 50
- 4.2 Product Quality — 51
- 4.3 Productivity — 51
- 4.4 Safety, Health and the Environment — 52
- 4.5 Summary — 53

5. The Outcomes of Filtration Processes
- 5.1 Filter Cake Outcomes — 57
  - 5.1.1 Cake Moisture — 57
  - 5.1.2 Cake Washing — 61
  - 5.1.3 Particle Breakage — 61
- 5.2 Filtrate Outcomes — 61
  - 5.2.1 Filtrate Clarity and Volume — 62
- 5.3 Slurry Outcomes — 63
- 5.4 Filtration Costs — 63
  - 5.4.1 Power Costs — 63
  - 5.4.2 Utilities: Air and Water — 64
  - 5.4.3 Consumables — 64
  - 5.4.4 Maintenance Costs — 64
  - 5.4.5 Operator Costs — 64
- 5.5 Examples of Filtration as a Part of a Process — 64
  - 5.5.1 Mineral Concentrate: Simple Dewatering — 64
  - 5.5.2 Alumina — 67
  - 5.5.3 Starch Washing and Dewatering — 71
- 5.6 Summary — 72

Part III
Filtration Process Success Factors

6. Slurry Filterability
- 6.1 The Nature of the Slurry to be Filtered — 76
- 6.2 Pretreatment of the Slurry — 76
  - 6.2.1 Density — 77
  - 6.2.2 Temperature — 77
  - 6.2.3 Additives — 78
- 6.3 Slurry Handling — 79
  - 6.3.1 Pumping — 79
  - 6.3.2 Storage/Suspension — 79
  - 6.3.3 Flow Control: Valves — 80
- 6.4 Summary — 80

Contents

7. Filter Design

7.1	Vacuum Filtration: Continuous	84
	7.1.1 Rotary Vacuum Drum Filter	84
	7.1.2 Rotary Vacuum Disc Filter	91
	7.1.3 Vacuum Belt Filter: Tray Type	94
	7.1.4 Vacuum Belt Filter: Rubber Belt Type	99
	7.1.5 Pan Filter	101
7.2	Pressure Filtration: Continuous	103
7.3	Pressure Filtration: Discontinuous	104
	7.3.1 Filter Press	104
	7.3.2 Tower Press	111
	7.3.3 Tube Press	115
	7.3.4 Candle Filters	117
	7.3.5 Spinning Disc Filters	120
	7.3.6 Leaf Filters	121
7.4	Centrifugal Filtration	121
	7.4.1 Batch Centrifuges	121
	7.4.2 Continuous Centrifuges	123
7.5	Summary	124

8. Filter Installation

8.1	Human Considerations	126
	8.1.1 Access: Walkways, Platforms, etc.	126
	8.1.2 Noise	126
	8.1.3 Lighting	126
	8.1.4 Ventilation and Extraction	127
	8.1.5 Flooring	127
8.2	Process Considerations	128
8.3	Slurry Systems	129
8.4	Cake Handling	129
8.5	Summary	129

9. Filter Cloth

9.1	Desired Outcomes	133
9.2	Cloth Design and Manufacture	133
	9.2.1 Materials	134
	9.2.2 Yarn Types	135
	9.2.3 Weave Types	136
	9.2.4 Postweaving Treatment	138
	9.2.5 Cloth Finishing and Fabrication	141
9.3	Cloth Support Grid	141
9.4	Cloth in Operation	142
	9.4.1 Cloth Failure	143
	9.4.2 Cloth Cleaning	144
	9.4.3 Cloth Repair	144
9.5	Summary	146

10. Filter Maintenance

10.1	Particular Issues with Filtration Equipment	148
10.2	Speed Versus Machine Sympathy	149
	10.2.1 Optimal Operating Regime for a Multiple-Filter Installation	149
10.3	Component Plant Trials	151
10.4	Summary	151

11. Filter Operation

11.1	Operational Choices	154
	11.1.1 Example: Trough-fed Filter Capacity Versus Speed	154
	11.1.2 Effect of Speed on Top-fed Filters	157
	11.1.3 Batch Pressure Filters	157
11.2	Summary	158

Part IV
Creating and Sustaining a Competitive Advantage

12. Process Testing

12.1	Test Equipment	162
	12.1.1 Laboratory Scale	162
	12.1.2 Pilot-Scale	163
12.2	Testing Program	164
12.3	Design of Experiments	164
12.4	Sampling	165
12.5	Example Method	165
12.6	Data Acquisition	165
12.7	Archiving Data	166
12.8	Cake Washing	166
12.9	Analysis	166
	12.9.1 Scaling Up	168
12.10	Summary	168

13. Getting the Most from Filtration Processes

13.1	Product Development and Process Design	170
13.2	Equipment Selection	170
	13.2.1 Principles of Equipment Selection	171
	13.2.2 The Process of Buying (and Selling)	172
	13.2.3 Second-hand Equipment	174
	13.2.4 Slurry and Application Questionnaire	175
	13.2.5 After Commissioning	175

13.3	Process Optimization	175
	13.3.1 Optimization Projects	177
	13.3.2 The Tower of Filtration Process Success Factors	178
13.4	Summary	179

Bibliography

Appendix A Useful Expressions

A.1	Slurry Relationships	183
A.2	Cake Relationships	185
A.3	Balances Across the Filter	185
A.4	Summary	187

Appendix B Flow Through a Growing Porous Filter Cake

B.1	Filtration Equation	189
	B.1.1 Pressure Drop over Cake: Δp_{cake}	190
	B.1.2 Pressure Drop over Medium: Δp_m	191
	B.1.3 Pressure Drop over the Cake and Medium	191
B.2	Filtration under Constant Pressure	192
B.3	Filtration with Constant Flow	194

Appendix C Forms and Templates

Appendix D Sample Test Method

Assessment of Slurry Filterability	199
Objective	199
Background	199
Apparatus	199
Set-up	200
Sample Collection and Preparation	201
Method	201
Analysis and Interpretation of Results	202

Index	203

Preface

This is the book that I looked for, but couldn't quite find, when I first started working in the field of solid–liquid filtration.

This is a time of great change. Given the way that the global process industries are being buffeted by the winds of globalization, economic uncertainty and environmental awareness, there has probably never been a better time to look for ways to:

- reduce costs
- improve product quality
- improve safety, health and environmental performance
- increase productivity.

At the same time, advances in filtration technology, especially in terms of cloth engineering, automation/control, filter aids and materials development, mean that there has probably never been a better time to be looking.

There is, rightly, a great deal of focus on reducing individual resource consumption, in particular energy and water, and it is to be applauded that many of us are looking to do this. Turning off the lights and computer monitors in chemical factory offices overnight is useful, but the impact on energy consumption can be dwarfed by relatively small improvements in the performance of a solid–liquid filtration process.

During the fifteen or so years that I have spent in this industry, many people have given freely of their ideas, inspiration and knowledge. In particular, I would like to mention Jaska Helsto, Timo Vartiainen, Ian Townsend, Rob Mclean, Doireann Funnell, Tim Ryan, Tim O'Connor, Tom O'Carroll, Brian Mawson, Kevin Schraden, Dirk Otto, Topi Karppanen, Luke Kirwan, Denise Croker, Lloyd Holliday, John Purdey, Richard Lydon and Antti Häkkinen, and thank them for their time spent in discussion.

A companian website for this book contains templates, simple software tools and other information for download. It can be found at: www.solid-liquid-filtration.com

Trevor Sparks
www.filter-ability.com

List of Figures

1.1 Aspects of processing competitiveness.
1.2 A simple description of the outcome of a filtration process.
1.3 A more detailed description of the outcome of a filtration process.
1.4 Ishikawa diagram of filtration performance success factors.
1.5 Success factors for filtration performance, shown as a wobbly tower.
1.6 The consequences of one success factor being seriously out of line.
1.7 The central ideas of the book.
2.1 Filter press in operation, early twentieth century (Wheal Martyn Museum, St Austell, UK).
2.2 Early rotary vacuum drum filter c. 1910 (FLSmidth GmbH).
2.3 Filter press in operation c. 1960s.
2.4 Global aluminum production since 1900 (source: www.world-aluminium.org/statistics).
3.1 One possible filtration (and washing) cycle.
3.2 The effect of certain stages on a simple dewatering application.
3.3 Potato starch particles (Salmela and Oja, 2006) and kaolin particles.
3.4 Single solid particles in a liquid, the liquid moving through a filter cloth.
3.5 A large number of solid particles in a liquid, the liquid moving through a filter cloth.
3.6 Large particles with a specific affinity for the cloth.
3.7 The filter cake, starting to grow.
3.8 Volume-based particle size distribution of two starches (note logarithmic x-axis).
3.9 Reduction in flow through a growing filter cake, under conditions of constant pressure.
3.10 Cumulative volume collected through a growing filter cake, under conditions of constant pressure.
3.11 Filter cake washing: replacement of mother liquid (dark) with wash liquid (light). Note the presence of dead space and shortcuts at the microscale.
3.12 Washing curves.
3.13 A simple counter-current washing system.
3.14 A reflux washing system.
3.15 Stages of air displacement.
3.16 Three examples of mesoscopic cake variation, showing the effect on final moisture (darker cake wetter, lighter drier). On the right-hand side, a crack in the cake is causing a very high flow of air, for little benefit in terms of drying (or washing).
3.17 Micrograph of diatomaceous earth, a filter aid (EP Minerals).
4.1 The four aspects of processing competitiveness.
5.1 Detailed description of the outcome of a filtration process.

List of Figures

- 5.2 Fuel consumption related to cake moisture.
- 5.3 A drying cost calculation in a spreadsheet (downloadable from www.solid-liquid-filtration.com).
- 5.4 Dryer capacity cost calculation in a spreadsheet (downloadable from www.solid-liquid-filtration.com).
- 5.5 Simplified flowsheet of a mining process.
- 5.6 A piece of copper ore, before and after grinding to liberate the copper-bearing part.
- 5.7 The principle of flotation.
- 5.8 Simplified alumina flowsheet.
- 7.1 A categorization of filter types.
- 7.2 General arrangement of a drum filter.
- 7.3 The control head: a key component in rotary vacuum filters.
- 7.4 Rotary vacuum drum filter, assembled ready for delivery (FLSmidth).
- 7.5 Rotary vacuum drum filter with press belt.
- 7.6 A variety of drum filter configurations.
- 7.7 General arrangement of a vacuum disc filter.
- 7.8 General arrangement of a tray-type belt filter, cloth tensioning and cloth tracking omitted. In reciprocating tray filters, the trays move, and are returned by the actuator shown. For stop–go filters, the trays are fixed.
- 7.9 Reciprocating tray belt filter.
- 7.10 Belt filter with multiple washing stages (BHS Sonthofen).
- 7.11 Belt filter cake discharge (Outotec Filters).
- 7.12 Belt filter, showing fume hood (Outotec Filters).
- 7.13 General arrangement of a rubber belt filter.
- 7.14 General arrangement of a pan filter.
- 7.15 Example of a belt filter in a pressurized enclosure.
- 7.16 Continuous pressure filter: all stages (feeding/filtration, washing and air drying) take place under pressure, giving a continuous cake discharge.
- 7.17 General arrangement of a side-bar filter press.
- 7.18 General arrangement of an overhead beam filter press.
- 7.19 A side-bar filter press (Outotec Filters).
- 7.20 Example of a plate and frame pack.
- 7.21 Detail of a diaphragm plate (r) with a recessed-chamber plate (l).
- 7.22 Diaphragm plate (with a chamber plate), showing the cloth support pips.
- 7.23 A chamber plate, showing the cloth support pips, stay-bosses and filtrate drainage.
- 7.24 A simple filter press cycle.
- 7.25 General arrangement of a single-cloth tower press, shown during cake discharge; cloth tensioning and cloth tracking omitted.
- 7.26 Tower press plate.
- 7.27 Tower press filtration cycle.
- 7.28 Cake discharge (Outotec Filters).
- 7.29 Tube press: (a) outer cylinder; (b) cross-section showing the central cloth-covered candle; (c) slurry feeding; (d) cake pressing; (e) cake discharge.

List of Figures

7.30 Tube press installation.
7.31 Candle filter: general arrangement.
7.32 Disc stack in a centrifugal discharge filter.
7.33 Schematic of a batch centrifuge. After the batch operations are completed, a paddle (not shown) moves in to direct the cake through the gap in the base of the basket.
8.1   An example of an installation (FLSmidth).
8.2   An example of a simple vacuum filter (continuous) installation.
9.1   Left to right: Monofilament and multifilament yarns (courtesy of Clear Edge Ltd).
9.2   Plain weave: the warp yarns are vertical and the weft go from right to left.
9.3   A 2/1 twill weave (over 2, under 1).
9.4   Satin weave (4:1).
9.5   Multilayer cloth, single side (cake on top).
9.6   A twill-woven cloth before and after calendering (courtesy of Clear Edge Ltd).
9.7   Coated cloth (courtesy of Clear Edge Ltd).
9.8   The cloth support grid. Left: cloth side; right: underside, which allows the filtrate to flow freely.
9.9   A small burr (the injection-molding point).
9.10 Pinhole in a multifilament cloth. This could have been caused by a stray sharp particle, a piece of weld splatter or a sharp edge on the filter equipment, e.g. the burr in 9.9.
9.11 Cloth wash sprays.
9.12 Filter press cloth cleaning system.
9.13 Before and after cleaning (courtesy of Clear Edge Ltd).
12.1 Laboratory test filter: this unit can operate as a pressure filter or a vacuum filter.
12.2 An example of a test data sheet. The spreadsheet can be used to give a mass balance across the process for solids and liquids to give an indication of the quality of the data.
13.1 Example of questionnaire.
13.2 The central ideas of the book.
13.3 A more detailed description of the outcome of a filtration process.
13.4 Tower of filtration process success factors.
A.1   Slurry broken down into its solid and liquid components.
A.2   Cake broken down into its solid and liquid components.
A.3   The filtration process.
B.1   Flow versus time through a growing filter cake.
B.2   Area under the curve.
B.3   Flow versus time through a growing filter cake.
C.1   Example of questionnaire.
C.2   Example of reference visit checklist.
C.3   Example of change assessment.
D.1   The filterability.xls spreadsheet.

LIST OF TABLES

9.1 General properties of some filter cloth materials
10.1 Operational scenarios
11.1 Vacuum drum filter scenarios

Chapter 1

Introduction

Chapter Outline
1.1 Scope 2
1.2 Summary 9

Creativity is just connecting things.

Steve Jobs, Wired Magazine, Feb 1996

There is a very good chance that the filtration processes on your production plant (or on paper, or even in your mind's eye, as a part of a plant design) can be made more successful and that this, in turn, will give your overall production process a competitive advantage. That is why this book exists. The central ideas in this book are:

1. That the competitiveness of a production process, relativey to your competitors, is determined by how it performs in terms of:
   - production cost
   - product quality
   - productivity
   - safety, health and the environment (SHE).
2. That the outcomes of filtration steps within that process can have a great impact on this competitiveness. (And that, whenever possible, it is a good idea to quantify this impact in currency terms – some of the numbers can be very significant and still surprise me.)
3. That a number of discrete success factors combine to determine the performance of a filtration process.
4. That by having a greater depth of knowledge about these success factors you can optimize your choices in:
   - process development
   - equipment selection/plant design
   - process optimization

and, thereby, create and sustain a competitive advantage.

The structure of the book follows this sequence of ideas.

The aim was not to produce a reference work on the *subject* of filtration; instead, I have tried to give some context to the subject as well as offer up some practical and applicable ideas that will lead directly to an improvement in your filtration processes. It can serve as an introduction for the uninitiated (giving suggestions for further reading for those wanting to deepen their knowledge), as a refresher for those who remember something about the subject from their process engineering education or, for someone with a deep theoretical knowledge of the subject, it can act to link their knowledge with more day-to-day challenges.

1.1 SCOPE

The field of filtration is large and wide, so we must be clear, from the beginning, about the scope of this book.[1] The book covers:

- Filtration of solid particles with typical particle sizes from 1 μm to 100 μm in a slurry with > 1% solids (weight by weight), for reasons that include:
    - removal of a small amount of solid matter from a liquor (i.e. polishing)
    - removal of liquid from a slurry (e.g. before transportation or evaporation)
    - removal of product in the liquid phase from a slurry (in this case, the solid can either be a waste or a by-product)
    - removal of a contaminant in the liquid phase from a solid product (the liquid may be recycled back into the process).
- Using slurry filtration equipment, including:
    - vacuum filters: disc, drum, belt and pan continuous filters and vacuum nutsch batch filters
    - pressure filters: chamber filters (filter press, tube press, automatic pressure filters), pressure vessel filters (candle, leaf and pressure nutsche)
    - centrifugal filters: peeler, inverting basket, pusher.
- Industrial applications, including:
    - pharmaceuticals: biotechnology, active pharmaceutical ingredients, fermentation broths, plant extracts, etc.
    - food/beverage starch: sugar, yeast, brewing, etc.
    - industrial minerals: kaolin, calcium carbonate, titanium dioxide, etc.
    - metals production: mining, metal refining, alumina, etc.
    - general process industry: bulk chemicals, waste treatment, polymers, etc.

Equations are used sparingly (where a picture can be used instead then it will be). This is because the emphasis is on improving process performance, rather than necessarily predicting it to a certain degree of accuracy. However, process engineers use spreadsheets for a great deal of their work and so calculations

---

[1] As a young salesman, attempting to sell large-scale slurry filtration equipment to chemical and mining companies – some of these machines could individually weigh up to 100 tonnes – I would still get calls from people: "Hello, I got your name from Yellow Pages, can you get me an oil filter for a 1971 Ford Capri?"

and models are described in more detail in the appendices and spreadsheets are made available for download from the website that accompanies this book.[2]

Throughout, I will refer to case examples to reinforce ideas. Where possible, these ideas and themes illustrated in these examples will be applicable to other cases or industries.

The book is written so that each of the central ideas is treated in turn. It is intended to be read as a whole, but can also be dipped into for a particular idea or piece of information. It is divided into four parts.

Part I serves to give some background to the field of solid–liquid filtration. Following a brief pseudohistorical review there is a background to the physical processes at play in an industrial filtration process. It will be noted, and explored, how the performance of full-scale filtration equipment (sometimes comprising more than 100 tonnes of machinery and hundreds of square meters of filtration area) is determined by how countless billions of microscopic particles interact with each other, and how a certain number of them interact with millions of pores in a filter cloth.

Part II will explore the first two of the central ideas, starting with:

1. ... the competitiveness of a production process, relative to your competitors, is determined by how it performs in terms of cost, quality, productivity and safety, health and the environment (SHE).

It makes sense, therefore, to assess our ideas for improving a process, and in particular a filtration process, within this framework, as shown in Figure 1.1. By framing competitiveness in this way, we will establish a useful tool for assessing the impact of filtration (or, as it happens, anything else) on an overall process. I believe that there should be space for Figure 1.1 on your office wall and that it should form the basis for your process key performance indicators.[3]

Next, we move onto:

2. ... the outcomes of filtration steps within that process can have a great impact on this competitiveness. (And, whenever possible, that it is a good idea to quantify this impact.)

Even the simplest type of filtration process can be characterized in terms of its inputs and outputs. Each of these can have a critical bearing on the success or otherwise of the process. Figure 1.2 shows the simplest form of a filtration process in which a slurry (a mixture of solids and liquids) is fed to a filtration device and filter cake (mostly solids) and filtrate (mostly liquid) are produced.

Now, imagine that the product of this process is drinking water. The amount of solid matter in the filtrate is clearly (it's sometimes difficult to avoid terrible puns) a critical quality factor that determines the success of the process.

---

[2] http://www.solid-liquid-filtration.com/
[3] In Part III, we will constantly refer back to these four aspects of competitiveness when weighing up alternatives for process development, equipment selection/plant design and process optimization.

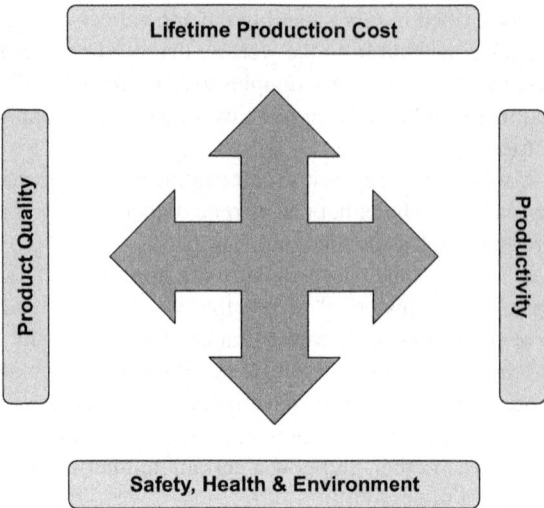

FIGURE 1.1 Aspects of processing competitiveness.

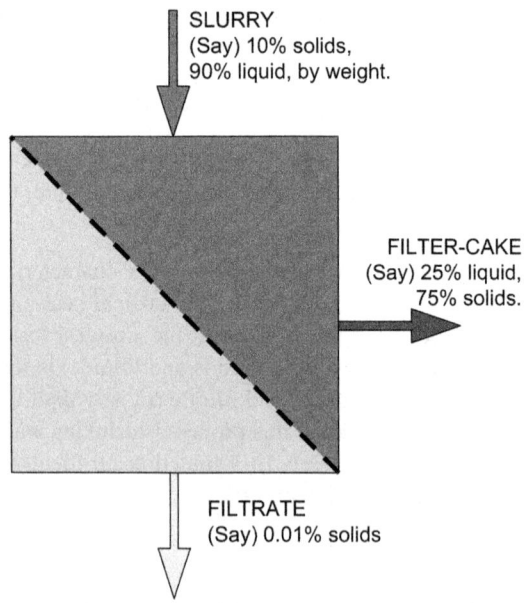

FIGURE 1.2 A simple description of the outcomes of a filtration process.

Chapter | 1 Introduction

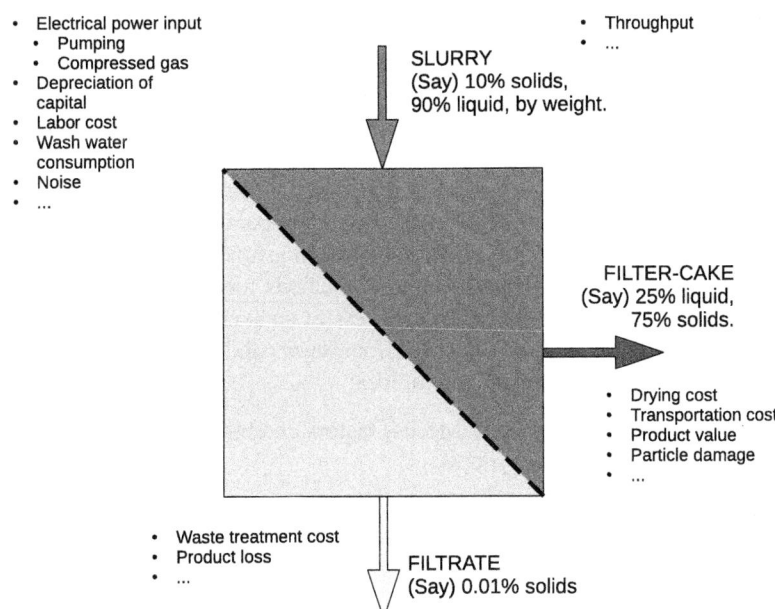

FIGURE 1.3 A more detailed description of the outcome of a filtration process.

Equally, if the solid material is a mineral ore, to be transported over a large distance to a metal refinery, then the amount of water can have a major impact upon the economics of transportation (a cost factor).[4]

An expanded, but still not necessarily complete, illustration of the outcome of the same process could look something like Figure 1.3. You can see that this list of outcomes looks beyond the simple outcome of the filtration process itself and notes things such as the cost of operation and noise (a serious consideration if your process is located near people's homes). It also identifies, alongside the filter cake, the cost of transportation and/or drying of the thermal filter cake, as well as something more obscure such as particle damage (which could be a critical product requirement in a pharmaceutical process).

Considering the filtrate outcome, this can have a bearing on the cost of waste treatment (which could be a kilometer from your filtration process), product losses (if the solid is your product) and a number of other possibilities.

A more complicated process, perhaps requiring filter-cake washing to remove a contaminant contained in the liquid phase (perhaps including three

---

[4] Say we have a production process producing 10 tonnes per hour (dry basis) of a mineral product, and the wet filter cake is transported 100 km to a customer in 40 tonne trucks. At 25% wt/wt cake moisture, approximately 2920 truck journeys are needed, while at 15% wt/wt, this number would reduce to approximately 2577 journeys, or 343 fewer. This is a number that can be used to build an argument for modifying the process; in this case, the ratio of the number of truckloads would be: $(1-25\%)/(1-15\%) = 0.88$: a spectacular improvement, and one that is entirely realistic.

stages of washing), would yield a larger number of other filtration process outcomes that must be considered. It can get quite complicated and there is a great deal to be gained in being methodical in the assessment of these outcomes and how they connect to your overall competitive position. I will recommend that you spend time producing your version of this figure.

The examples above are rather simple, but I hope that they illustrate the usefulness of spending some time exploring the impact that outcomes of filtration processes have on an overall process, and in a structured way. This part of the book includes simple calculations and methods for quantifying the impact of filtration outcomes on the relative success of an overall production process. Some of these can also be downloaded from www.solid-liquid-filtration.com

In Part III, we take the next central idea:

3. ... that a number of discrete success factors combine to determine the performance of a filtration process.

In fact, six discrete factors are identified and each of them is given its own chapter. They are shown as an Ishikawa (or root-cause) diagram in Figure 1.4.

However, while each of these factors makes a contribution to the overall performance, they each hold an effective veto as well. For example, a process to separate a slurry that is eminently filterable, using a well-made and appropriate piece of filter machinery that is well-installed, maintained and operated, will still fail if the filter cloth is inappropriate, damaged or at the end of its useful life; this can apply equally to any of the other individual factors. So, it is better to imagine that each of these factors is supporting the overall filtration process performance as a tower (Figure 1.5).

This tower looks rather unstable, as it should, because if any one of these factors is poorly selected or has deviated significantly from the norm, then the result can be disastrous (Figure 1.6).

Each of the chapters in Part III discusses one of the six success factors. Topics covered will vary from the technical and fact based (e.g. slurry distribution over a filter element) through to the more human (e.g. teamwork in filtration

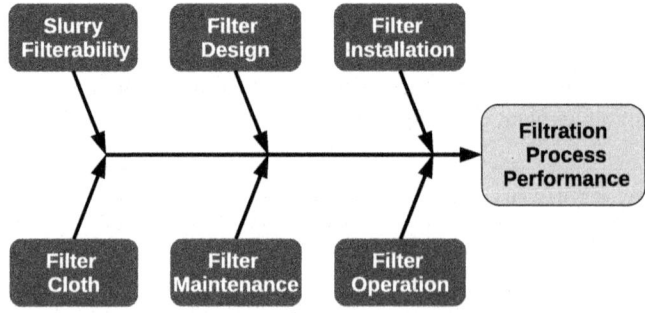

FIGURE 1.4 Ishikawa diagram of filtration performance success factors.

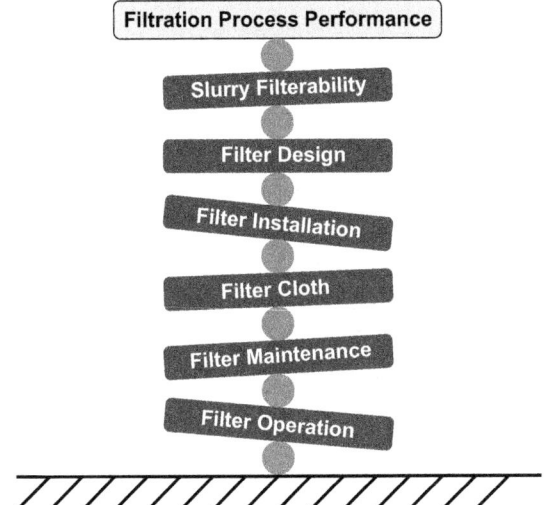

FIGURE 1.5 Success factors for filtration performance, shown as a wobbly tower.

FIGURE 1.6 The consequences of one success factor being seriously out of line.

plant management, creation of a safe and as pleasant as possible working environment).

Part IV takes the last of the central ideas:

4. ... by having a greater depth of knowledge about the [success] factors (shown in Figure 1.5) you can optimize your choices in process development, equipment selection/ plant design and process optimization and thereby create and sustain a competitive advantage.

We will focus on the things that can be done to create and sustain the competitive advantage that we are looking for. This will incorporate discussion of laboratory- and pilot-scale testing, monitoring of processes (process analytical technologies) and buying (and selling) of filtration equipment. The aim of this section is to reinforce and apply the other central ideas.

Finally, the appendices contain some mathematical background to the physical processes involved as well as provide sample calculations, templates and other sources of information.

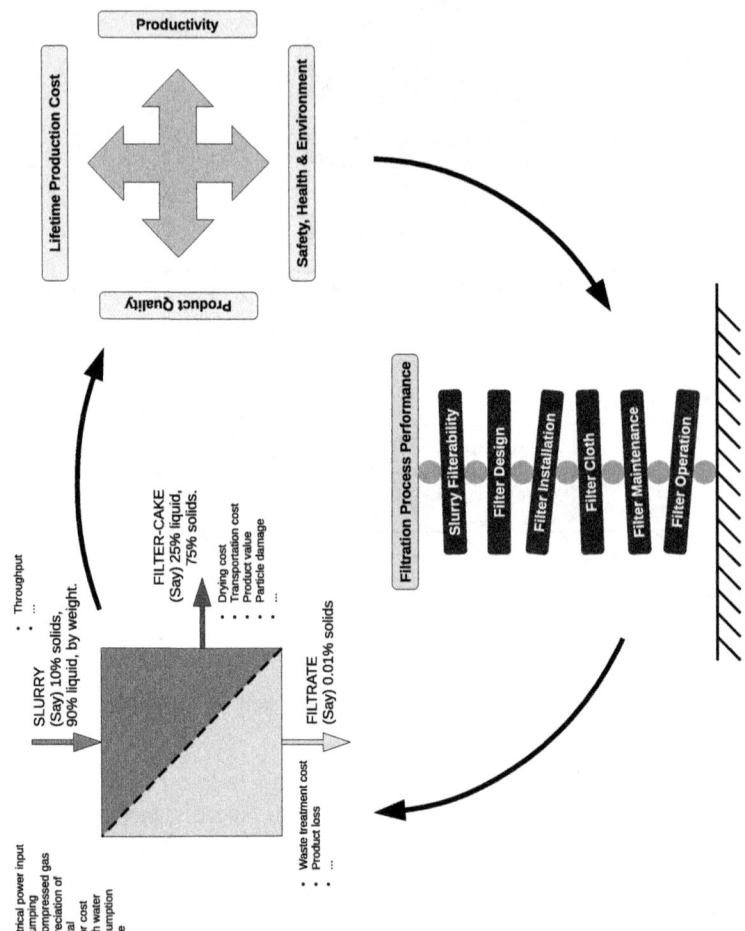

FIGURE 1.7 The central ideas of the book.

1.2 SUMMARY

The aim of this book is to help you to create and sustain a competitive advantage for your process, by optimizing your solid–liquid filtration choices in process development, equipment selection/plant design and process optimization.

The book explores four central ideas and develops them:

1. That the competitiveness of a production process, relative to its competitors, is determined by how it performs in terms of cost, quality, productivity, and safety, health and the environment (SHE).
2. That the outcomes of filtration steps within that process can have a great impact on this competitiveness.
3. That there are a number of discrete factors combining to determine filtration performance.
4. That by having a greater depth of knowledge about these factors you can optimize your choices in process design, equipment selection and process optimization and, thereby, create and sustain a competitive advantage.

By considering your filtration choices in the knowledge of how outcomes will affect the success of your process and through having deeper knowledge of the factors that combine to deliver these outcomes, you will make better choices.

This is summarized and illustrated in Figure 1.7.

Part I

Background

Chapter 2	History of Filtration	13
Chapter 3	Physical Phenomena	21

Chapter 2

History of Filtration

Chapter Outline
2.1 Origins   13
2.2 Industrial Revolution
     and Onwards   14
2.3 Recent History   16
2.4 Current Trends   17
   2.4.1 The Shape of the Industry
          Today   18
2.5 Summary   19

This chapter will review the origins of filtration and look at what has driven, and what has enabled, filtration technology, and its supporting industry, to develop to the present day.

2.1 ORIGINS

Humankind has been engaged in solid–liquid separation since before recorded history. Someone probably realized that the water she had just scooped out from a muddy pool would be better to drink if she waited for the bits to settle before sipping the clear liquid from the top. It must only have been a matter of time before somebody refined this idea and noticed that you could get clear water slightly more quickly by passing the muddy water though a bundle of reeds, a simple piece of cloth or a bag of sand. And so filtration, as an engineering discipline, would have been born – in this case out of the need to improve the productivity and quality of a product (drinking water).

Early recorded evidence of filtration includes a carving in the tomb of Pharaoh Amenhotep II showing a water purification device and, a few centuries later, Hippocrates wrote (in 400 BC):

[Rain waters] are the best of waters, but they require to be boiled and strained; for otherwise they have a bad smell, and occasion hoarseness and thickness of the voice to those who drink them.

So there is no question that humans understood the importance of drinking water clarity for good health, even without necessarily knowing about microorganisms or other microscopic contamination.

From that time forward, filtration was used in brewing, viticulture and food-oil production. There are numerous cider and olive presses in folk museums throughout Europe. The fruit, perhaps in woven bags, would be pressed in these contraptions and relatively clear liquid would emerge.

As the need developed for dyes (to color clothing, for example) as well as other basic chemical products, so the technologies used for production became more sophisticated. Filtration is mentioned in the Stockholm and Leyden X Papyrus, written around AD 300, which includes recipes for making several simple chemical products.

The word filter itself was in use in the English language from as early as the fourteenth century and is related to the Latin word for felt.[1]

2.2 INDUSTRIAL REVOLUTION AND ONWARDS

With the Industrial Revolution came the means to intensify and control power to previously unimaginable intensities. It also meant that people flocked from the countryside to live in towns and cities and work in factories. The resulting high population densities brought the need for water and wastewater treatment while, luckily, advances in technology made them possible.[2] The industrialization of manufacturing gave the means to produce everything from metals to ceramics at a lower cost and on a massive scale.

Other engineering devices and technologies, many of them useful to solid–liquid separation, originated or were developed during this time. Without centrifugal pumps, vacuum pumps, industrial textiles, metal casting and welding, very few modern filtration technologies would be viable.

Some important developments in filtration and separation came from the production of kaolin, or china clay. Huge quantities of this material, used in the production of tableware, were discovered in Cornwall, in the UK, in the late eighteenth century.[3] The clay would be extracted and classified using a wet process; the majority of the material mined (the waste, or spoil) would form the piles that can be seen in that part of the world even today. The challenge was to remove the water from the product so that it would have the right consistency to form cups, plates, toilet basins and tiles. Initially, this was done by a combination of settling (leaving the clay for several days and decanting off the water) and thermal evaporation.[4] However, this whole process was accelerated enormously by the introduction of filtration. The Wheal Martyn Museum in Cornwall contains a filter press from 1911. Figure 2.1 shows a similar filter in operation from around that time. It looks similar to a modern side-bar filter press (Figure 7.17).

---

[1] The German word for felt (*Filzen*) comes from the same root.
[2] In Britain, the first standards for water quality were introduced in the 1852 Metropolis Water Act, which specifically required the passing of drinking water through sand-bed filters (following their successful development in the 1830s).
[3] The British Empire was largely an exercise in getting tea, coffee and sugar to Britons from India and the West Indies; fine tableware completed the picture.
[4] Usually this meant raking clay around floors heated by fires underneath; it was both labor-and energy-intensive.

Chapter | 2 History of Filtration

FIGURE 2.1 Filter press in operation, early twentieth century (Wheal Martyn Museum, St Austell, UK).

FIGURE 2.2 Early rotary vacuum drum filter *c*. 1910 (FLSmidth GmbH).

In Germany, and other parts of Europe, a number of developments took place, many of these in parallel with the development of coal-fired power stations and the metal-refining industries. An example of an early, open-ended, rotary vacuum drum filter is given in Figure 2.2.

2.3 RECENT HISTORY

More recently, say over the past four decades, globalization and further rapid population growth have driven the need for process performance to even higher levels. At the same time, the technology enablers have continued to develop:

- cloth technology: materials and weaving/finishing technology
- manufacturing and design technology, leading to previously unimaginable levels of precision
- materials technology
- programmable logic control and automation.

Figure 2.3 shows a more recent filter press in operation. In contrast with Figure 2.1, this image shows developments in machine technology: this unit is

FIGURE 2.3 Filter press in operation *c.* 1960s (Wheal Martyn Museum, St Austell, UK).

an overhead beam filter press (Figure 7.18), giving better access to the plates/cloths, and each plate is mounted on a roller so that opening is easier. There are also developments in personal protective equipment: this operator is wearing long-sleeved overalls and gloves!

During the 1960s and 1970s, many people became more aware of the impact that our actions have on the environment, both locally and globally. At this time there were also several oil price shocks and, even if you do not subscribe to the overwhelming consensus on climate change, it was self-evident that energy efficiency was desirable and that it served companies well to seek it out.

The energy crises during the 1970s prompted the need for energy efficiency in the world's process industries and this led many to seek more effective solid–liquid filtration systems. This, in turn, led to increasing adoption of pressure filtration throughout the world's mining industry, since higher pressures tend to give higher specific filtration capacities and lower cake moistures compared with vacuum filtration.

The unfortunate production of acid rain – produced when sulfur dioxide produced from burning coal, for example, reacts with water in the atmosphere – has been virtually eliminated in large parts of the world. Flue-gas desulfurization involves wet scrubbing of ignition gasses through milk of lime to produce solid calcium sulfate (gypsum). Filtration, and belt filtration in particular (see Sections 7.1.3 and 7.1.4), can be used to wash the precipitated gypsum so well that it can be used as a building material.

2.4 CURRENT TRENDS

Today, urbanization is taking place at a much faster pace and on a larger scale than ever before and this brings the same issues as in the Industrial Revolution (water, waste treatment, healthcare, etc.), but this time on an unprecedented level. While the sustainability of China's economic model is up for debate, the scale of what has happened to date is not.[5]

In the early part of the twenty-first century, the need for clean water, energy, food supply, medicines, metals and plastics is growing at an extraordinary pace. Figure 2.4 shows the growth in production of aluminum, as more and more uses have been found for a metal that, a little over 100 years ago, was considered to be a precious metal. It is likely that production has at least doubled between your birth and 2010, some readers may have been alive while production increased ten-fold.

This figure is particularly relevant because each million tonnes of alumina (an essential step on the way to metallic aluminum) requires several hundreds of square meters of filtration capacity, usually drum, disc, pan and leaf filters (see Section 5.5.2).

---

[5] Whereas the number of large-scale filter presses produced in Europe per year have been numbered in the hundreds, there are individual factories in China today producing hundreds per week.

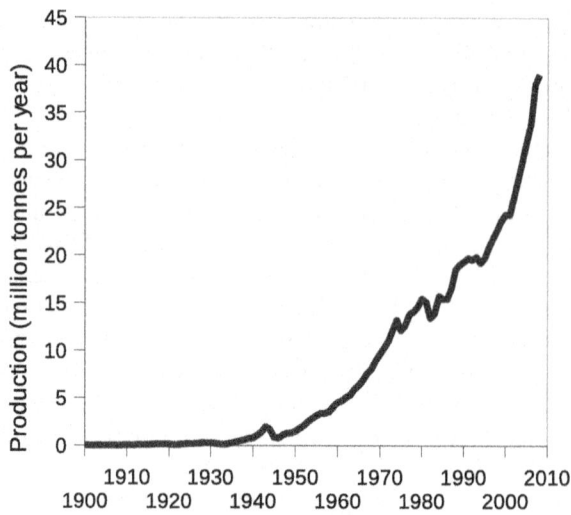
FIGURE 2.4 Global aluminum production since 1900 (source: www.world-aluminium.org/statistics).

2.4.1 The Shape of the Industry Today

There is a significant industry supplying equipment of the type covered in this book, with combined sales of new equipment of more than €2 billion per year. These suppliers typically fall into four categories:

1. General fabrication workshops that can manufacture generic filtration equipment from drawings or by copying and adapting existing equipment.
2. Specialist filter machinery manufacturers with a small number of filtration technologies available.
3. Larger companies that have a range of filtration technologies, often gained through acquiring companies in category 2.
4. Even larger companies that supply a wide range of equipment to the process and mining industries (e.g. grinding mills, conveyors, flotation columns, thickeners, clarifiers, calciners). These companies are aiming to provide a complete process to their customers.

Some companies progress through these categories, starting as general workshops and, many years later, becoming absorbed into much larger groups. It is my opinion that the degree of process expertise generally peaks with suppliers in category 2 (and possibly 3) and that a dilution of knowledge and expertise can accompany the addition of other technologies and products.

Today, the industries supplying solid–liquid filtration equipment and media are highly consolidated. Whereas, a couple of decades ago, the industry had a number of suppliers with, essentially, one product, today there are a number of filtration houses that can supply a wide range of filters or filter cloths; and this is constantly changing.

2.5 SUMMARY

The use of solid–liquid separation and filtration started with small-scale water purification and, later, the preindustrial production of foods, dyes, beers and wines. During this time, most filtration duties could be fulfilled using pots and simple cloths.

With the Industrial Revolution came the need for, and the means to produce, large-scale solid–liquid filtration.

The development in filtration technology has been both driven by need and enabled by the availability of technologies. In particular, developments in cloth, manufacturing design and precision, material science and control/automation have produced great advances in the past few decades.

Today, the drivers of the process industries – population growth and urbanization – are stronger than ever, as are the demands placed upon these industries in terms of environmental protection.

Chapter 3

Physical Phenomena

Chapter Outline

3.1 Basic Narrative Of a Filtration Process   23
3.2 Step-by-step Narrative   25
  3.2.1 The Nature of a Solid–Liquid Suspension   25
  3.2.2 Particles and the Filter Medium, the Very Early Stages of Cake Formation   26
  3.2.3 Cake Growth   29
  3.2.4 Cake Pressing   33
  3.2.5 Cake Washing   33
    3.2.5.1 Counter-Current Washing   37
    3.2.5.2 Reflux Washing   38
  3.2.6 Cake Draining   38
  3.2.7 Air Drying   39
    3.2.7.1 Hot Gas Drying   41
3.3 Other Notable Phenomena and Things to Look out for   41
  3.3.1 Variations in Cake Properties   41
    3.3.1.1 Microscopic Variations   41
    3.3.1.2 Macroscopic and Mesoscopic Variations   41
  3.3.2 Rapidly Settling Slurries   42
  3.3.3 Compressible Filter Cakes   43
  3.3.4 Migration of Fines   43
  3.3.5 Precoat and Body Feed   43
3.4 Summary   **44**

Imagination is more important than knowledge.
    Albert Einstein, *Philadelphia Saturday Evening Post,* October 26th, 1929

Within the scope of this book, a filtration process converts a mixture of solids and liquids into a separate filter cake and liquid (filtrate). There can be many reasons for doing this, but they all stem from the fact that the separated components are more valuable (or perhaps less costly) when they are apart than when they are together. Some examples are:

- To remove solids from a liquid: The liquid might contain a valuable product in solution and the solid is a worthless, or even harmful, residue. An example is a pregnant liquor on a metal refinery.
- To remove liquid from solids: The solid might be a mined metal ore that is transported across the world to a refinery; the liquid is simply water and

transporting it would be a waste of shipping (as well as a potential hazard if the system reslurries and sloshes about in the ship's hold.)
- To recover both solids and liquids: In a cellulosic bioethanol process both the sugar-rich solution (to be used for fermentation to ethanol) and solid lignin (to be burnt as a high-grade fuel) are valuable, but only if they are apart; the presence of one, in with the other, spoils its effectiveness.

In the perfect world, this separation would be absolute: there would be no liquid at all remaining with the solids and no solids at all in the liquid. It would also require little effort. In reality, however, all filtration processes are somewhat imperfect, and there will be moisture in the cake and solids in the filtrate.

The filtration process may simply be a single step along a longer separation process, perhaps preceded by gravity or centrifugal thickening and followed by thermal evaporation. In addition, the amount of effort required, and therefore expense, for filtration can be very significant.

A liter of a typical industrial slurry contains tens of billions of individual particles suspended in liquid and the outcome of a filtration process that we observe at our scale (and that we can measure using flow meters, scales, moisture analysis, turbidity meters or by counting the number of trucks that leave a site, etc.) is the aggregate result of countless billions of interactions between these microscopic particles, in the presence of a motive force for filtration and a filter medium.

It is crucial, then, to try to make sense of what is happening at this microscopic scale in these processes in order to understand their outcome. In fact, most problem-solving situations which I have did not involve come across, mathematical modeling or precise calculations, but rather thinking about *what could be happening*, coming up with an idea for improvement and then testing it out (preferably at laboratory or bucket scale first; see Chapter 12). Inference and thought experiments can be every bit as powerful as mathematical modeling of a situation. This approach will serve you equally well whether you are in process development (which may be years from full-scale production) or if you have a more immediate problem (say only eight hours to process, as well as you can, a batch of material that was badly precipitated overnight).

The huge variation in filtration applications means that it is difficult to be specific in the following sections: you may be interested in separating coarse, fragile crystals of an active pharmaceutical ingredient from an organic solvent or fine particles of crushed and ground mineral from water. The relevance of the particular phenomena discussed in this chapter will vary according to the application, and there may be other things happening that apply to any particular case.

The illustrations below are not meant to be accurate representations of real physical situations (the drawings would be too messy). They are instead meant to illustrate a point or an idea.[1]

---

[1] Recently, several studies have used microscopy and computational modeling to look in detail at these processes (e.g. Lu, 2006) and, in time, these should become more and more useful, in the same way that computational fluid dynamics has become a useful tool for fluid flow devices.

Chapter | 3 Physical Phenomena

First, we will review a filtration process on a rather basic and superficial level, to give some context to the descriptions that follow, and then the rest of the chapter will explore the phenomena that occur at the microscopic scale, the domain of particles and filter cloth pores, in more detail.

I strongly recommend that you take some time to think about some of these descriptions to see whether there are any ideas that you can generate to get more from your own solid–liquid filtration processes.

3.1 BASIC NARRATIVE OF A FILTRATION PROCESS

Figure 3.1 shows a typical filtration cycle. This can be either a cycle that occurs as a batch (as in a Büchner filter test in the laboratory) or the sequential stages on a continuous filtration device, like a belt or rotary drum filter.

The players in this process cycle are:

- Slurry: The solid–liquid suspension, i.e. the mixture of solids and liquids that we wish to separate.
- Motive force for filtration: This can be:
  - a pressure difference, provided by, for example:
    - a slurry pump
    - applying an overpressure (e.g. from an air compressor)
    - applying a vacuum to the system so that atmospheric pressure pressing down on the slurry provides the motive force
    - a piston or an inflated bladder
  - an acceleration field (gravity or centrifugal).
- Filter cloth: A medium with pores that will allow liquid to pass but will retain solids. In this book, the vast majority will be woven polymer cloths (polypropylene, polyester, nylon) but it can also be a woven wire or non-woven felt. Chapter 9 is devoted to a discussion of filter cloths.
- Filter cake: The dense structure of solids and retained liquid that accumulate on the filter cloth.
- Mother liquid: The liquid from the slurry retained in the filter cake. This may be further removed by drainage or gas displacement. It may also be replaced with another liquid (cake washing).
- Filtrate: The liquid that passes through the filtration cloth (and filter cake).
- Gas (optional): More liquid may be removed from the cake by passing gas (usually air, but sometimes other gases, such as nitrogen) through the cake.
- Wash liquid (optional): In order to improve product recovery (if held in the liquid) or reduce contamination, the mother liquid in the cake may be washed out of the cake.

Figure 3.1 only shows one possible simple filtration cycle. From this basic cycle, there are many other combinations. For example, it may incorporate a number of washing stages, even using different wash liquids (perhaps even wash filtrate from previous wash stages in a counter-current washing scheme).

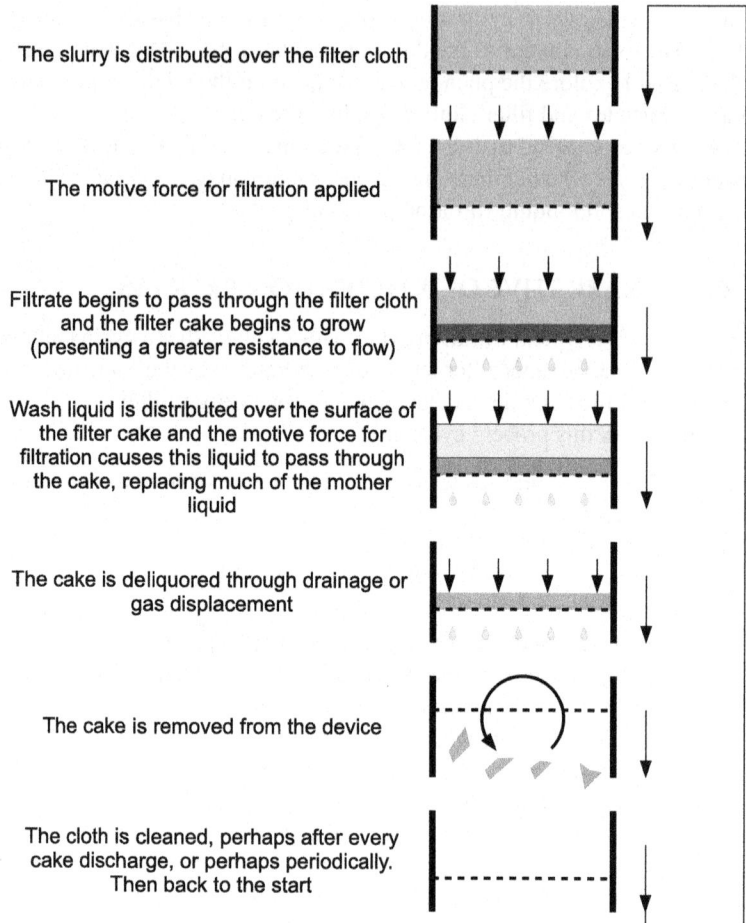

FIGURE 3.1 One possible filtration (and washing) cycle.

It may also include a stage in which the slurry/cake is squeezed with a piston or bladder (not shown in Figure 3.1).

As stated previously, these individual steps can occur either one after another in a batch filter, say a laboratory Büchner, or simultaneously at different places on a continuous filter, for example belt or rotary vacuum drum filters.

Each of the steps performs a part of the overall required duty, as shown in Figure 3.2, in which the rate of liquid removal in each of the three stages reduces with time. This figure also illustrates the idea that different slurries may respond differently during the stages, for example, some slurries/filter cakes give up a greater proportion of their liquid during air blowing.

The rest of this chapter will look in more detail at the steps, and at the nature of the slurry, filter cake, filtrate and the microscopic phenomena that determine the performance of solid–liquid filtration processes.

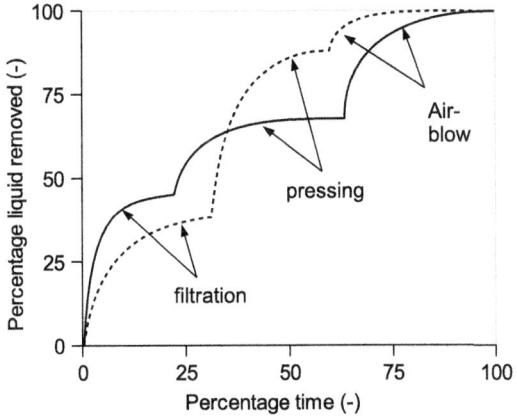

FIGURE 3.2 The effect of certain stages on a simple dewatering application.

3.2 STEP-BY-STEP NARRATIVE

3.2.1 The Nature of a Solid–Liquid Suspension

In a typical industrial solid–liquid suspension, or slurry, the countless billions of particles have a rather loose interaction with each another. They may interact with those particles that are very near, but will not affect (or transmit force to) those much more than one or two particle sizes away. The liquid in the suspension, as a fluid, does transmit force and is subject to normal fluid dynamics (continuity, viscous drag, turbulence, transmission of pressure, etc.).

Figure 3.3 shows two different types of solid particle from industrial slurries. There is a marked contrast in the size and shape: large and rounded starch particles, and much smaller, plate-like and angular kaolin particles.

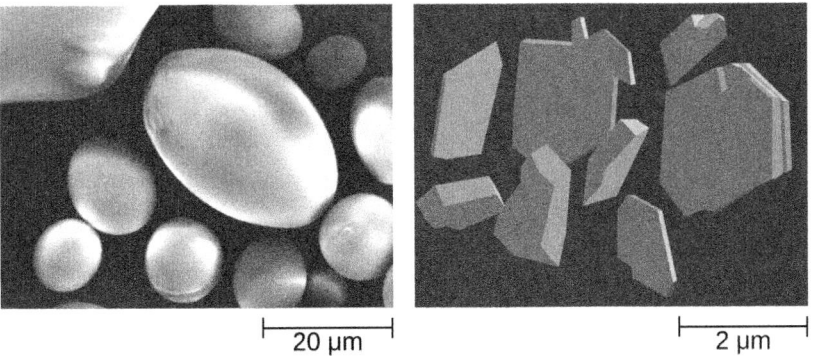

FIGURE 3.3 Potato starch particles (Salmela and Oja, 2006) and kaolin particles.

The nature of interactions between particles can vary tremendously according to their size, size distribution and shape. Particles in suspension in the size range of this book (1–100 μm) carry surface electrical charges, and these result in repulsive and attractive forces between the particles. These forces become increasingly important, relative to inertial forces, the smaller the particle size. The effect of these surface charges also depends critically upon the nature of the liquid (pH or the presence of a coagulant or flocculant). In some situations, the particles may become tightly bound together in microscopic clumps (coagulated). In any case, it can be misleading to think of the particles simply stacking together like snooker balls, as the interactions may not involve physical contact.

In the absence of gravity (e.g. on a space station's wastewater treatment plant), these particles would move about, bashing into each other, perhaps staying together, perhaps not, until ultimately the energy of the system was dissipated and it came to a rest. On Earth (and to a lesser extent on a lunar station's wastewater treatment plant), particles that are more dense than the liquid will settle to the bottom, in most industrial situations over a period of several minutes, hours or even days.

In the presence of a motive force for filtration (a pump, pressure difference, or centrifugal or gravity field), the slurry will move as a liquid.

3.2.2 Particles and the Filter Medium, the Very Early Stages of Cake Formation

We start with the simplest arrangement possible: single solid particles contained in a liquid that is flowing through the pores of a cloth; this may be because of applied pressure difference or an acceleration field (gravity or centrifugal) (Figure 3.4).

Particle A, on the left, which could be a potato starch grain from Figure 3.3, is clearly larger than the pore and cannot pass through the filter cloth. It is dragged towards the filter cloth and captured by blocking (or sieving or straining). The particle on the right, perhaps a kaolin particle, is somewhat smaller than the

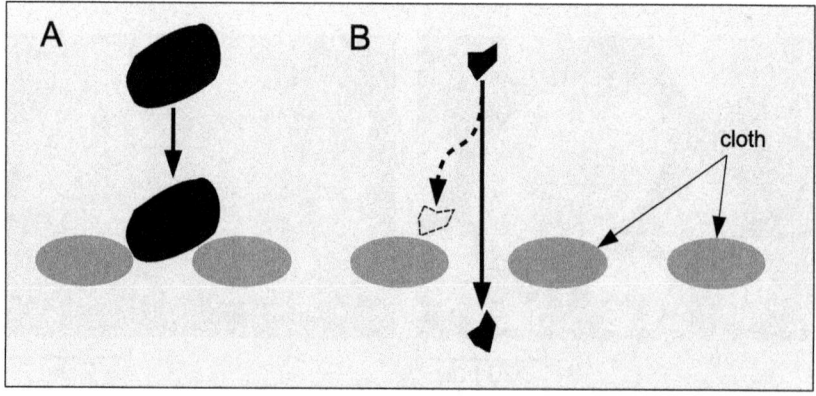

FIGURE 3.4 Single solid particles in a liquid, the liquid moving through a filter cloth.

Chapter | 3 Physical Phenomena

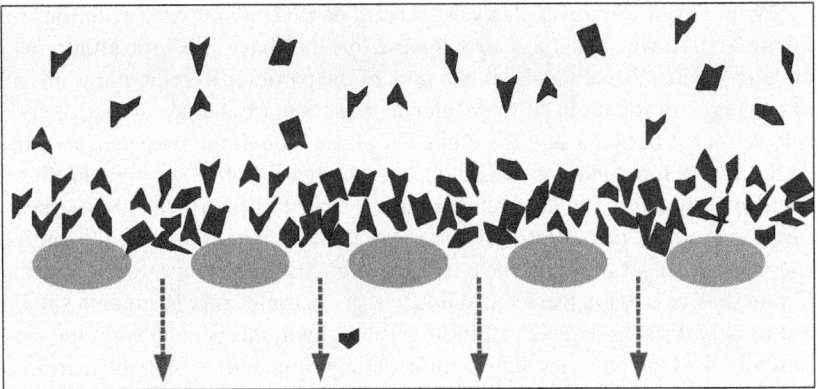

FIGURE 3.5 A large number of solid particles in a liquid, the liquid moving through a filter cloth.

pores in the filter cloth (which is often the case in industrial filtration applications); so, in the absence of any reason not to, in other words not counting any attractive forces between the cloth and the particle, it will probably pass straight through the opening. Let's just say, for now, that this type of particle would normally pass through this type of opening 95 times out of 100.

Next, we can imagine the same filter cloth and liquid, but this time with a very large number of small particles suspended in the liquid, each one being of a similar size to the single particle from before (Figure 3.5). This time, all of these particles are jostling to find their way through the pore and there is a good chance that many of them will jam together to form bridge-like structures over the pore (especially given the repulsive and attractive forces that may exist between the particles). So, even though each individual particle would, in isolation, normally pass straight through the pores in the filter cloth, in larger numbers, they have a greater tendency not to. By extension of this point, while filtering, say, 1 kg of solids from 100 liters of liquid may produce a cloudy filtrate, filtering the same quantity of identical solids from 10 liters of slurry (with the same local fluid velocities, etc.) will probably produce a clearer filtrate.

In any case, there is a relationship between the concentration of solids in the slurry and the clarity of filtrate, and this can be seen on many filtration plants.

Now, if the particles are sharp or abrasive, then any particles that pass through the pores in the cloth may cause damage to the yarns of the cloth. So, it is reasonable to assume that there would be a relationship between the amount of solids in the filtrate and the amount of damage caused to the cloth – and therefore the lifetime of the cloth. Given the relationship between slurry density and solids in filtrate, then, higher slurry densities may lead to better cloth lifetimes.[2]

---

[2] On some industrial minerals plants, a filter shutdown may be automatically triggered by a drop in measured slurry density, perhaps caused by a disruption to an upstream thickener or the hydrocyclones. An hour or two of lost production while the thickening process is restored to health is seen as preferable to a whole shift lost replacing all of the filter cloths.

Figure 3.6 illustrates a slurry with large particles (compared with those in Figure 3.5). However, in this case, suppose that they have a specific affinity with the filter cloth. (While this is shown here as the particles forming plugs in each of the pores in the cloth, it may refer to a particularly strong attractive force between large particles and the cloth.) So, even though the particles are much larger than before, and you would expect a higher filtration throughput, this is not always the case. The important point here is that filtration can often surprise. I have performed laboratory tests on slurries that, when looking at the particle and particle shape, I would have expected to filter well, only to find out that filtrate flow is very poor indeed. While a larger particle size distribution should, and usually does, have a better filtration throughput, this is not always the case. In some cases the particles almost immediately form an impenetrable barrier to flow on the cloth. (The type of cloth could have a bearing, so it is worth trying another.) There may be other factors that could be adjusted to assist in improving the slurry–cloth interaction that could have a positive bearing on filtration.

A number of other factors will determine how these early stages of cake formation at the filter cloth proceed.

- The speed of flow: Looking back at Figures 3.4 and 3.5, it is apparent that the more slowly the small particle is moving, the more likely it is to be trapped by the cloth (or by the structures that are beginning to form near the cloth). In practical terms, it may be that gentle flow during the early stages of cake formation would pay dividends.
- Viscosity of liquid: Liquid passing through the system will impart drag forces on particles, and if this drag is more significant than the forces holding back a particle, then it will pass through the cloth. Since this drag is highly dependent upon viscosity and since the viscosity of a liquid is very sensitive to temperature, you may see improved filtrate clarity (fewer particles being dragged through the cloth) with a higher temperature.

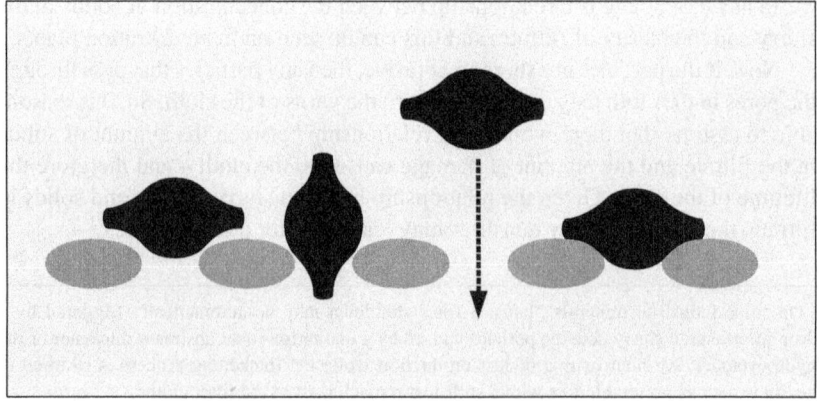

FIGURE 3.6 Large particles with a specific affinity for the cloth.

- Gravity forces: In an acceleration field (gravity or centrifugal), the particles may settle towards (and even onto) the cloth at the same time as filtrate flows. In some centrifuges, or with a very coarse slurry, this settling may be almost instantaneous, so the whole filtration process will take place with the cake already formed. This cake may have different properties over its depth, with coarse particles close to the filter cloth and finer particles near the cake surface.
- Surface electrostatic charge: As the particles jostle for position and form bridges over the pores of the cloth, the relative repulsive and attractive forces will have a major influence on how they settle into position. This can be seen at the large scale by the difference in filterability of, for example, kaolin over a pH range or with the addition of coagulants and flocculants.
- Particle shape: Most methods for assessing the characteristics of particles in a slurry focus on the particle size distribution, but particle shape, as well as surface charges, can affect how the particles come to settle together.

3.2.3 Cake Growth

Once the early stages of the filter cake formation are over, with bridge structures, for example, over pores in the filter cloth, slurry will continue to report to the newly formed cake surface. This filter cake is now acting as a filter medium itself and, crucially, the cake is usually far more effective at trapping particles than the cloth itself since the apparent pore size presented to the approaching particles is much smaller, as represented in Figure 3.7. Once

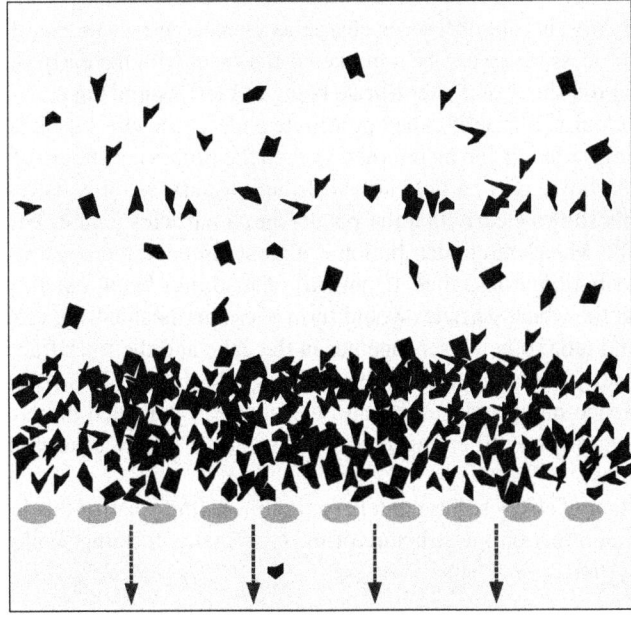

FIGURE 3.7 The filter cake, starting to grow.

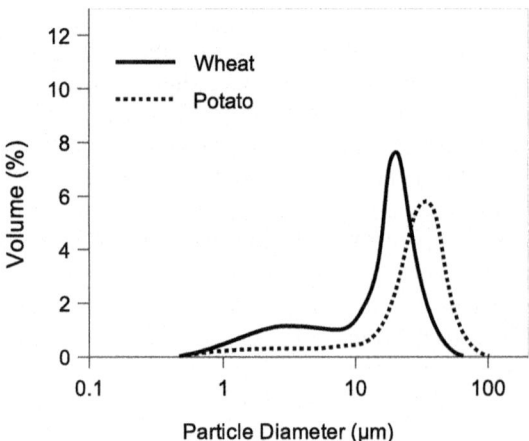

FIGURE 3.8 Volume-based particle size distribution of two starches (note logarithmic x-axis).

a particle enters the filter cake it has passed an event horizon, or point of no return, and as long as slurry continues to follow from behind, it is unlikely to break free back into the slurry above the cake, unless the cake is disrupted somehow.

Some particles, especially very fine particles, may migrate forward through the crowd of particles to arrive at the pores in the cloth. However, as the cake thickness increases, the chances of these particles getting through reduce. At the macroscale, you may see some cloudy filtrate at the beginning of a filtration cycle. This filtrate often becomes clearer as the cake grows thicker. Depending upon the process, there can be a marked difference, with the early stage filtrate resembling the slurry and later filtrate being as clear as gin. On many processes the early, cloudy, filtrate is called prefiltrate and, if recovery of solids or filtrate clarity is important, it can be returned back to the process upstream of the filter.[3] Figure 3.8 shows two particle size distribution curves, for potato and wheat starch. This shows clearly that the potato starch particles tend to be larger, but also that the wheat starch distribution is almost bimodal: there is a second peak to the distribution at less than 10 pm. All other things being equal, you would expect that the wheat particles would form a less permeable filter cake (the particles would tend to be closer together in the cake and the finer fraction would tend to occupy pores in the cake between larger particles). This is borne out by the experience on an industrial scale: wheat starch is indeed more difficult to filter, giving a lower filter capacity and, usually, a higher cake moisture (see later sections).

The effect of slurry solids content (sometimes simply called the density) will also have an effect on the structure of the cake that is forming. Following very

---

[3] This may not always be permitted in certain processes covered by regulatory requirements.

Chapter | 3   Physical Phenomena

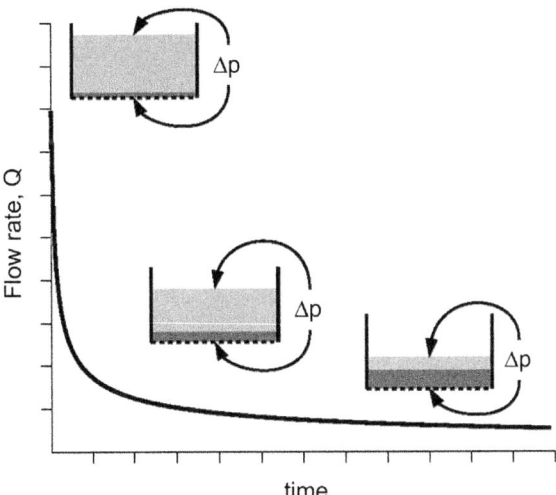

FIGURE 3.9   Reduction in flow through a growing filter cake, under conditions of constant pressure.

similar reasoning to before (contrasting the case of a single particle to many particles arriving at a pore in a filter cloth), any gaps in the surface of a filter cake will tend to be filled by isolated particles but many particles will tend to compete for this space and form more bridge-like structures around the gap. The result is a more structured, open, permeable cake that presents less resistance to flow and will therefore have a higher liquid throughput (for the same quantity of solids and the same dewatering force).

So, a higher slurry density will tend to produce clearer filtrate and have a higher permeability: there can be very good reasons indeed for prethickening as much as possible.

If you watch a filtration process happen in the laboratory, say at constant vacuum, it is obvious that the rate of flow decreases as the cycle proceeds and the cake grows (Figure 3.9). As the cake grows, the resistance to the flow of liquid through the cake increases.

Figure 3.10 shows the amount of filtrate collected from a similar test. For the great majority of filtration applications within the scope of this book, the volume of filtrate collected, under conditions of constant pressure, is approximately proportional to the square root of the time.

$$V \propto \sqrt{\text{time}} \tag{3.1}$$

In other words, to receive double the amount of filtrate (and double the amount of filter cake) takes about four times as long. This relationship is shown and developed more in Appendix B, but it is of crucial importance to the relationship between capacity and the speed or filtration-step time of a filter. You can test this relationship yourself in the laboratory or even on the plant.

Some filtration processes operate on the basis of constant flow (either if the flow rate is controlled or if a constant flow pump is used). In this case, the

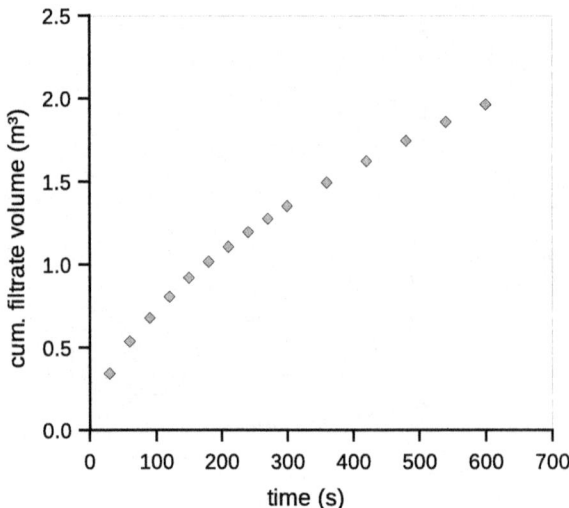

FIGURE 3.10 Cumulative volume collected through a growing filter cake, under conditions of constant pressure.

pressure drop across the filter cloth and cake will start small and grow. In many cases, the pressure drop is proportional to the cake thickness (and therefore the amount of filtrate collected). Since the amount of filtrate collected per unit time is constant, you will often see a linear increase in pressure drop. An exception to this is when the filter cake is compressible (see Section 3.3.3).

As before, when the slurry was beginning to form a cake on the filter cloth, similar factors affect the nature of the cake as it grows:

- Speed of flow: The higher the speed of the liquid passing through the cake, the more likely that particles will pass through the cake. For this reason, polishing filters (removing small quantities of solids from large volumes of liquid) are often sized according to the speed of liquid. This speed may be surprisingly low, in the order of 1 cm per minute.
- Viscosity of liquid: Higher viscosity will tend to drag more particles free of the bonds of the cake. In addition, higher viscosity, and higher drag forces, will bring particles to the cake with greater force and tend to produce a more compact, and therefore less permeable, cake structure.
- Surface electrostatic charge: As the particles jostle for position and form bridges over the gaps in the surface of the cake, the relative repulsive and attractive forces will have a major influence on how they settle into position and, therefore, the structure and permeability of the cake.

Overall, the filter cake is a precious thing; it protects the filter cloth from potential damage, clarifies the filtrate and it should be treated with care. Most good filtration technologists and test engineers spend a great deal of their time thinking about what could be happening in their filter cakes, imagining the way that the particles form in the cake, perhaps the way that fines migrate through the cake or the way that coarse particles may settle quickly relative to the time taken to form the cake.

It is this stage that largely determines the capacity of the filtration process (in terms of the amount of slurry processed). Once the cake has been formed, a number of possible process steps can follow, in all manner of different possible sequences.

3.2.4 Cake Pressing

In this case, a force is applied to the filter cake by a piston or a pressurized bladder (sometimes called a diaphragm or even membrane[4]).

In many filters, this cake compression can be greater than 16 bar (or even above 150 bar) and, in some applications, almost no cake formation occurs before the high force of a pressing bladder is applied to the slurry.

As the bladder comes under pressure, this will be resisted by an increase in pore pressure in the cake (or slurry above the cake). Eventually, the pore pressure in the cake will drop, as the pressure is borne by the structure of the cake itself. After this point, there is little to be gained, at least in terms of liquid removal, from further pressing. If you are able to measure the pressure of the cake, then this can be a useful trigger to control a process.

Some fragile particles within a filter cake may be damaged by the high forces (or the high rate of change of force) of the pressing diaphragm as the cake consolidates and the particles are ground together. This can affect the final product particle size distribution and even produce fine particles that may emerge in the filtrate later in the cycle (e.g. during washing or air drying).

Equally, some fragile cake structures can also be destroyed by cake pressing. I have seen cases where a reasonable cake is formed by pumping slurry into a chamber, only for it all to be pressed through the filter cloth a few moments later.

The cake pressing stage sometimes acts to prepare the cake for optimal washing or air-drying performance (see below) by producing a compact, uniform structure. As with many situations in filtration, the effectiveness of a cake pressing stage can only really be determined through testing. However, tell-tale signs that some form of cake pressing is needed might be severe cake cracking or cake shrinkage during air drying during a laboratory test without pressing (see Section 3.3.1).

3.2.5 Cake Washing

Where a higher degree of separation between solids and the liquid in the slurry is needed, the formed (and perhaps also pressed) cake can be washed. Examples include:

---

[4] Not to be confused with membrane filter media in microfiltration or ultrafiltration.

- Fermentation broth: The mother liquid may contain the primary product (say an antibiotic intermediate). The recovery of this liquid should be maximized. The solids may have another application, if you are able to remove the antibiotic intermediate.
- Modified corn starch: The solids are the primary product, but are only fit for human consumption if the amount of modifying agent (e.g. an inorganic acid) is reduced to below a certain limit.
- Metal refinery residue: A precious metal may be dissolved in the liquid phase, with the solids being a less valuable residue that will be disposed of or processed to produce a less valuable product.

The washing stage can follow on from cake formation (simply by switching the slurry feed over to wash liquid) or a preformed cake can be flooded with wash liquid. In some cases, the washing stage can follow cake pressing or, sometimes, even gas drying. Often the best, and possibly the only, way to find the optimal operating sequence for a filtration process is through testing.

In the ideal case, the wash liquid will pass through the cake as a uniform front, replacing all of the liquid as it proceeds, so that all of the mother liquid is removed and replaced with precisely the same volume of wash liquid. Figure 3.11 illustrates the idea of this replacement washing. The lighter colored wash liquid passes through the filter cake, replacing the contents of the pores as it goes.

The figure hints at the fact that some of the routes through the cake may be blocked for the wash liquid while other routes may be comparatively easy. In reality, a number of mother–liquid volumes will be needed. Some of the main reasons for this deviation from the ideal include:

- Microscopic effects; for example:
  - porous particles that hold the mother liquid within themselves
  - clumps of agglomerated or flocculated particles that trap the mother liquid
  - a chemical affinity between solids and the contaminant
- Mesoscopic effects; for example:
  - cracks in the filter cake that allow wash liquid to short-circuit
  - gaps that form around the edge of filter cakes if they shrink
  - thick or thin regions in the filter cake (see Section 3.3.1)
- Macroscopic effects; for example:
  - Filtration equipment characteristics: for example, the wash liquid distribution over the cake may not be ideal, or, in the case of multichamber filters, may be poorly distributed between these chambers.

One possible indication of washing performance is the concentration of the contaminant in the filtrate (and wash filtrate) emerging from the filter and this can be measured, for example, with an online filtrate conductivity probe.

Chapter | 3 Physical Phenomena

a) Before Washing.

b) Washing. Note the wash-front and remaining mother liquid.

c) Washed.

FIGURE 3.11 Filter cake washing: replacement of mother liquid (dark) with wash liquid (light). Note the presence of dead space and shortcuts at the microscale.

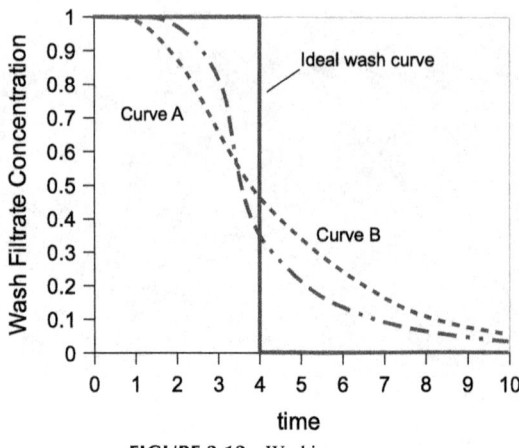

FIGURE 3.12 Washing curves.

Figure 3.12 shows some examples of wash filtrate concentration versus time. In the ideal case, the wash front proceeds through the cake, all the while pushing mother liquid out of the bottom of the cake (and without any mixing or molecular diffusion) in this case, after a certain point in time, all of the mother liquid will have been displaced and, suddenly, pure wash liquid will emerge as the filtrate. This is shown as the ideal wash curve (the time for this to take place also includes an "equipment time", for example, the time taken for a feed manifold to be filled with wash liquid.

The other curves in Figure 3.12 represent deviations from this ideal mixing, short-circuiting and diffusion. The early reduction in wash filtrate concentration in these curves could be a sign of short-circuiting (either through a single chamber, a large gap or many smaller cracks in the cake). The long tail on a wash curve can represent the washing out through mixing or diffusion of the last stubborn remains of the mother liquor in a cake. In reality, this very slowly approaches zero and in some extreme cases large quantities of wash liquid will be needed.[5].

If the cake has already been air dried or drained, then the wash curve will be rather different as the wash liquid first refills the pores in the cake then flushes out the mother liquid that remained after deliquoring (not shown in Figure 3.12).

In a minority of cases, where physical flushing of the cake is difficult, some success may come from a pause in the process, with the cake flooded with wash liquor, to allow molecular diffusion to reduce the concentration in these stubborn pockets of resistance.

---

[5]The most extreme case that I have seen is in a speciality ink pigment – more than 100 liters of deionized wash water per kilogram of solid product – although there are surely even more demanding applications.

Chapter | 3 Physical Phenomena

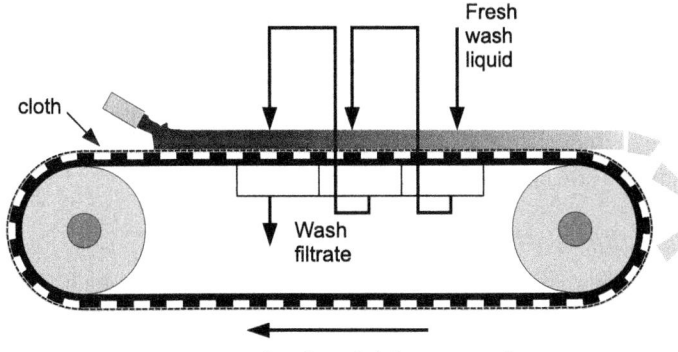

FIGURE 3.13 A simple counter-current washing system.

Finding optimal washing configurations often requires a great deal of imagination and creativity, the best attributes of process development and filter test engineers.

3.2.5.1 Counter-Current Washing

The overall effectiveness of a washing operation *can* be improved with a counter-current washing scheme and, at the same time, the amount of wash liquid needed can be reduced. The basic idea is to reuse wash filtrate (produced using fresh wash liquid) to wash the cake before it arrives at the last washing point and so on, as many times as you like, until the benefits are outweighed by the cost of extra stages or it simply becomes impractical to add more stages.

There are several reasons why counter-current washing should improve a washing process. First, the more passes of any wash liquid through the cake, the better (even if it includes some mother liquid). Second, if there is a significant difference between the characteristics of the mother liquor and wash liquid (e.g. viscosity) then the wash liquid may be ineffective in removing the mother liquid. (Imagine water being used to remove golden syrup from a filter cake: it will probably not be very effective; however, slightly diluted syrup will be more effective, and so on.)

On some filters (notably horizontal belt and pan filters; see Sections 7.13, 7.1.4 and 7.1.5) the counter-current wash can be continuous. Figure 3.13 shows a three-stage counter-current wash system on a vacuum belt filter (Section 7.1.4).[6] The fresh wash liquid is added at the end of the belt and washes cake that has just been washed with the final filtrate which, in turn, has been washed with stronger wash filtrate. While there may be a significant benefit in

---

[6] In fact, belt filters can be used with many more stages; potentially with no limit although, practically, up to about ten.

using counter-current washing and a noticeable improvement when going from two to three, there will be a diminishing return in going from, say, eight to nine.

A counter-current washing scheme could also be built using three filters in series, with the wash filtrate from the last unit going to the previous one, or from a batch filter, using an arrangement of tanks, pipes, valves and pumps (or beakers if you are doing it in the laboratory).

3.2.5.2 Reflux Washing

Another possible cake washing scheme is reflux washing, in which a large recirculating loop of wash liquid passes through the cake, with a smaller bleed-in of fresh wash liquid and take-out of wash filtrate (Figure 3.14).

3.2.6 Cake Draining

Some industrial slurries produce cakes that are free draining under gravity. Others will drain in the exaggerated "gravity" of a centrifuge. The resulting liquid content in the cake is the result of the competition between the acceleration field trying to remove the liquid and the tendency of the cake to retain the liquid: surface tension between liquid and particles is a major factor, as illustrated in the next paragraph.

Soak three canvas bags, containing 1 kg of coarse gravel, 1 kg of sand and 1 kg of fine clay, in a tank of water (it could take days for the day to become fully saturated). If you then remove the bags one at a time, the water will drain away from the gravel bag almost immediately, leaving a thin film of water on the grains and other small pools held by surface tension at the points where the grains are touching. The final moisture content of the bag will be rather low. Most of the water will drain from the bag containing sand after a few seconds, although the grains will also be coated in a film of water and there will be liquid held by surface tension between many of the particles. The total moisture in the bag will be higher, given that the total surface area of the grains in the bag will be much higher (so there will be more held in

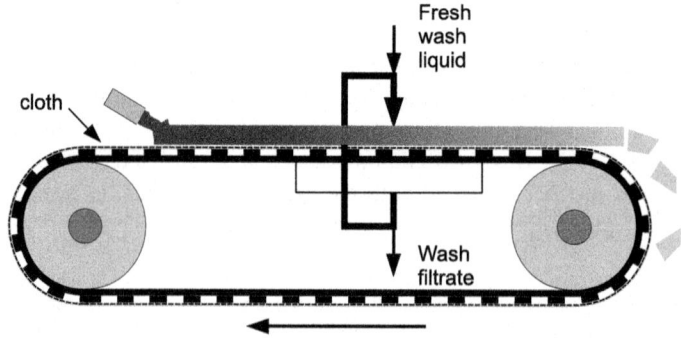

FIGURE 3.14 A reflux washing system.

the films coating each grain) and there are more points where particles are close together to hold liquid. Finally, almost no liquid will emerge from the bag of clay.

In a similar way, in the final spin in a centrifuge, the final result depends upon the balance of centrifugal forces trying to fling the liquid outward and the surface tension trying to keep the liquid on the particles.

3.2.7 Air Drying

Air drying is commonly used to deliquor filter cakes.[7] The ultimate form of this would be one-to-one replacement of all liquid in the cake with air, in the same way that wash liquid would replace all mother liquid in the idealized washing case. In this scenario, with a perfectly uniform filter cake, the gas displacement would pass through the cake as a uniform front. However, in reality, cake drying deviates from the ideal case for many reasons, many of them similar to those listed for cake washing; for example, variations in the cake, or over a filter machine and liquid trapped in pores in the cake.

If you monitor the removal of liquid during this stage of a filtration cycle, there is usually a significant flow in the early stages of air drying as the "easy" moisture is cleared (the air flow can also clear out any liquid that is lying in pools in filtrate trays or pipes, adding to this early rush). Thereafter, there is a reduction in the rate of liquid removal (accompanied by an increase in volumetric flow rate air and, therefore, of cost) until there comes a point when there is nothing to be gained in prolonging the air-drying stage.

The net effect of air drying is a combination of direct displacement (particularly in the early stages), the gradual scouring out of pockets of moisture and finally evaporation (Figure 3.15). As with cake draining, the final outcome reflects the balance of these factors against surface tension and closed pores that will tend to keep moisture within the cake. When looking at the relative costs of solid–liquid filtration, the air-drying stage is usually the most expensive in terms of cost per unit volume of liquid removed (because of the high flow of air from air compressors or through vacuum pumps). However, provided that the time used (and therefore volume of air) is reasonable, this is still usually more cost effective than thermal evaporation.

There may be a worsening in filtrate clarity during air drying as fine particles find new routes to migrate through the cake. As the cake dries, there may be high localized air velocities and these may cause some erosion of certain points in the filter, especially if the air is transporting a significant amount of solid matter.

---

[7] Other gases can be used, e.g. nitrogen, but for simplicity we will refer to gas drying throughout this section as air drying.

a) Before air drying.

b) Drying.

c) After drying.

FIGURE 3.15 Stages of air displacement.

3.2.7.1 Hot Gas Drying

The removal of moisture can be further encouraged using hot gas (or even steam) to remove moisture directly in the cake through evaporation as the hot gas passes through the system. This may also heat up the filter cake so that some moisture continues to evaporate after the cake has been discharged.

However, you should take care that the components of the filter are able to withstand high temperatures and sudden changes in temperature. Some filter cloths and rubber materials (e.g. pressing bladders) may be vulnerable.

3.3 OTHER NOTABLE PHENOMENA AND THINGS TO LOOK OUT FOR

The stages of a filtration cycle described above are reasonably universal and occur in most industrial filtration equipment in some combination or sequence. Some deviations and special cases will be introduced in this section.

3.3.1 Variations in Cake Properties

As discussed in the sections on cake washing and air drying, in an ideal situation the filter cake will be perfectly uniform and homogeneous, presenting precisely the same resistance to the flow of filtrate, wash liquid and drying air. This situation does not exist in industrial solid–liquid filtration equipment. The best that you can hope for is that these variations are not too significant.

Variations in cake properties can be microscopic, macroscopic (on the scale of the filter equipment) or mesoscopic (on the scale of the filter cake, and often visible).

3.3.1.1 Microscopic Variations

Coagulating, bringing together, particles to form larger effective particles, can bring a significant benefit to gravity thickening or gravity clarifying operations. These coagulated groups of particles, if they are tightly bound, will also tend to form a filter cake with a higher overall permeability and, therefore, capacity. However, they can also form an effective cage around any mother liquor contained in the cake. If this mother liquor is of great value, or is a contaminant, then the coagulated particles, which assisted in gravity settling upstream and in the basic filter ability of the slurry, may no longer be a good thing.

Optimization of coagulation and flocculation can be a rather inexact science, and trial and (plenty of) error may be the best approach to optimizing. In some applications, the slurry may be deliberately sheared to break down its structure before being filtered.

3.3.1.2 Macroscopic and Mesoscopic Variations

A thinner filter cake will offer less resistance to liquid flow, and therefore any regions of a cake that are thinner than the surrounding cake will become a shortcut for wash liquid or drying air. The opposite is the case for thicker regions of the cake. These variations can have a number of different causes:

- Filter equipment design:
  - poor overall design, e.g. the slurry or wash liquid feed velocities are too high and the cake is washed away
  - unsuitable design for this particular duty (e.g. the slurry is very fast filtering and the filter design is for a slower filtering application).
- Variations in cloth condition:
  - pin holes leading to thin regions (sinks)
  - cloth blinding in some regions but not others.
- Operation out of design range:
  - overfilling or underfilling
  - too high a filter speed.
- Cake cracking/shrinkage, creating channels through which air or wash liquid can rush without meeting, and therefore not removing, mother liquid (Figure 3.16). Cracking can occur as the overall volume of the cake reduces (because of the removal of moisture), creating tensions in the cake. Cracks release this tension.

3.3.2 Rapidly Settling Slurries

As discussed in Section 3.2.2, coarse, dense particles may settle rapidly, compared with the time taken to form a filter cake. As a result, very rapidly settling slurries (usually containing particles larger than 100 μm) tend to be filtered on top-fed filters (e.g. horizontal belt or pan filters). Rapidly settling slurries may be unsuitable for many filters because of the difficulty in forming a uniform filter cake.

Trough-fed filters (where the filtration surface is downward facing or vertical during cake formation) may need an agitator within the slurry trough to keep the slurry suspended.

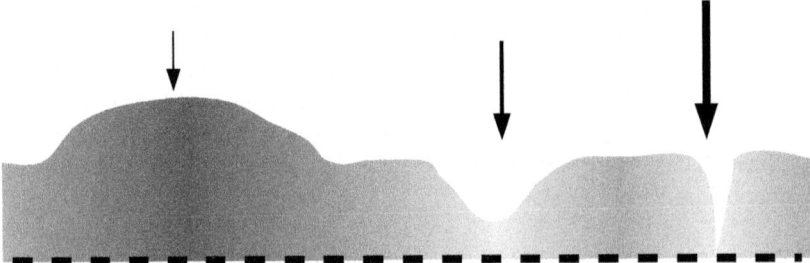

FIGURE 3.16 Three examples of mesoscopic cake variation, showing the effect on final moisture (darker cake wetter, lighter drier). On the right-hand side, a crack in the cake is causing a very high flow of air, for little benefit in terms of drying (or washing).

3.3.3 Compressible Filter Cakes

Until now, the descriptions of phenomena in the filter cake have assumed that the particles are held within a coherent and incompressible structure. If these particles are very large and rigid (imagine coarse sand with most of the particles in physical contact), this structure would not change in any meaningful way if the applied pressure increased. The particles would already be as close to each other as possible. Increasing the pressure will simply increase the rate of liquid flow.

However, in industrial cake filtration, the properties of many filter cakes vary according to the applied pressure. Crucially, higher pressures may produce a more compact cake, with lower permeability (or higher resistance to flow). This means that you will not receive all of the extra capacity that you would expect with higher filtration or pressing pressures. In some extreme situations (highly compressible cakes), the capacity of a filter may fall at higher filtration pressures (or higher spin speeds on centrifuges).

Cake compressibility is treated in more depth in, for example, Rushton et al. (2000) and Wakeman and Tarleton (2005a).

3.3.4 Migration of Fines

Throughout these descriptions of what could be happening within filter cakes, it has mostly been assumed that the filter cake has constant properties throughout its depth; in other words, that each new layer of cake that it laid down has the same porosity and composition as the layer below: the cake is homogeneous. However, one can easily imagine that very fine particles will tend to migrate through the cake during cake formation. If this is occurring, and if filtrate clarity is an issue, then reducing the particle inertia and the drag on the particles may help to alleviate this (by either reducing the velocity of liquid through the filter cloth/cake, at the expense of capacity, or reducing the viscous drag on the particles by maintaining a higher temperature).

3.3.5 Precoat and Body Feed

In some cases, filtration of a slurry in its native form may not be possible or practical. For example, deformable or gel-like particles may form an impenetrable coating on any filter cloth (which is essentially a two-dimensional array of pores). In these cases, a viable filtration process can be created by first forming a bed of porous material, or precoat. Figure 3.17 shows one possible precoat material, diatomaceous earth, but other fibrous or porous materials such as perlite (porous volcanic rock) or cellulose are also widely used.[8]

---

[8] Diatomaceous earth is composed of the fossilized remains of microalgae that lived between 3 and 65 million years ago. It is a highly porous material: try picking up a bag and you will feel the difference from, say, a same-sized bag of cement.

FIGURE 3.17 Micrograph of diatomaceous earth, a filter aid (EP Minerals).

This bed of precoat is now a three-dimensional array of pores, channels and surfaces, presenting far more opportunities for filtrate to flow, while unfortunately contaminating the solid material.

The other motivation for using a precoat, even if the solids in the slurry are not especially gelatinous, is to create an extremely clear filtrate. Precoat filtration is used widely in the beverage industry (activated carbon is used as a precoat that also removes color and fine haze from a liquid).

If the surface of the precoat bed becomes blocked then some filter designs allow for this layer to be shaved off, creating a fresh filtration surface (see Section 7.1.1.11).

Instead of being used to form a precoat, these same filter aids can be added to the slurry as a body feed to bulk up the slurry and ensure that any cake formed has a greater permeability. In some cases, the filter aids can be used for both precoat and body feed.

3.4 SUMMARY

The outcomes of filtration processes are, ultimately, determined by how countless billions of particles interact with each other and, some of them, with a filter medium or cloth. Very often, more progress can be made by thinking a process through:

- coming up with ideas for what could be happening at the macroscale, mesoscale and microscale in the filter cakes.

and then

- testing this idea and drawing conclusions.

Inference can be every bit as powerful as mathematical modeling of the situation. However, various mathematical models of filter cake formation, compression, washing and air drying exist. This means that you can make informed decisions when looking to develop or improve a filtration process.

Part II

Competitive Advantage

Chapter 4	Competitiveness in Processing	49
Chapter 5	The Outcomes of Filtration Processes	55

Chapter 4

Competitiveness in Processing

Chapter Outline
4.1 Production Cost   50
4.2 Product Quality   51
4.3 Productivity   51
4.4 Safety, Health and the Environment   52
4.5 Summary   53

My mother said to me, "If you become a soldier, you'll be a general; if you become a monk, you'll end up as the Pope." Instead, I became a painter and wound up as Picasso.

Pablo Picasso

All production processes operate in a competitive business environment. The days of protected, localized, markets are gone and your closest competitor (in commercial terms) could be many thousands of miles away. The purpose of this short chapter is to set out a framework for competitiveness, so that the implications of any choices that you make have a context. The importance of solid–liquid filtration process performance is not explicitly connected to this framework, but it will be throughout the rest of the book.

After a number of years looking around, meeting people, reading annual reports and attending conferences in the process industries, it is hard to come to any other conclusion than that the success of a process can be defined completely in four dimensions:

- production cost
- product quality
- safety, health and the environment (SHE)
- productivity.

If you are better than all of your competitors in all of these dimensions (assuming that there is a market for your product and that your salespeople have at least a basic level of competence) then you are doing very well. However, in most real cases, there is room for improvement, somewhere.

While some of these dimensions may influence another (e.g. a pollution incident may lead to a fine from the Environment Agency which in turn will affect production cost), it is very hard to find a simpler, or better, way of categorizing the requirements put upon a process; and I have really tried. Across the world, whether they are making computer chips, cars or chemicals, production managers need to keep these four dimensions of success in mind at all times. Most of them do, even if not in as simple a framework as this.

It is also likely that optimizing in one dimension could compromise another: quality may suffer if costs are reduced, or achieving the highest standards in SHE may have some impact on costs.

At a strategic level, a company can make choices within this framework; for example, they may choose to be the lowest cost supplier (and win market share on price) while perhaps not being able to serve the section of the market that demands the highest quality or purity. Other companies may choose to target the premium, high-quality, end of the market for a period to establish a good brand or reputation, before later starting larger scale production, all the time trying to maintain the hard-won reputation.

Very simply, if you are able to find a risk-free way to improve one dimension of success, without affecting any of the others, then you will do it. I challenge you not to, "No, I don't want to cut production costs/improve product quality/improve SHE/increase productivity, because …".

The following sections list and discuss some of the important considerations within these four aspects of competitiveness.

4.1 PRODUCTION COST

Here, we must include every cost that cannot be avoided in making a unit of product (a car, a tonne of paint, 1000 headache tablets …):

- raw materials
- capital (depreciated over a suitable period)
- energy
- labor
- transportation
- utility costs
- water
- effluent treatment
- overheads and administration.

Assuming that you are able to match your competitors in the other three dimensions of success, then you can now either make more margin at the same selling price as them, or gain more market share by reducing your price.

Most of the items on this list are self-explanatory, but are often neglected. An interesting idea is the use of waste as a raw material: some companies pay several dollars per tonne to dispose of their waste, and would happily give it for next to nothing to anyone who is willing to take it away. There are even precious metals producers who are reprocessing waste heaps from old mines, where most of the costs of mining were incurred decades earlier.

Companies that are able to make use of lower cost raw materials (perhaps they are lower cost because they are contaminated or of lower potency) usually have to develop their process to compensate.

4.2 PRODUCT QUALITY

There are two sides to consider here. First, have you managed to reach the basic requirements of a certain market? It may be possible that your product is suitable for an industrial use (e.g. in wallpaper paste); but, with a small improvement, you could sell it for a food use (e.g. as an ingredient in a soup). In normal circumstances, the soup manufacturer will pay more for your product than the wallpaper paste manufacturer.

Second, if you meet the entry requirements, how does your quality compare to that of your competitors? Perhaps the price level for industrial application of a product is around $50 per tonnes, but for a food application it is around $100. However, a food-grade product that has a better taste or better applicability (maybe it stirs in more quickly or at a lower temperature) could command $105, especially if your sales and marketing forces are able to communicate the benefits.

A further crucial aspect of quality is consistency. It is not really good enough to produce a food that is the most delicious and wholesome that money can buy for 99% of the time, but that for the remaining 1% might may cause terrible stomach upsets. Most risk-averse consumers would prefer to buy food that is pretty good, for 100% of the time, in terms of:

- purity
- taste
- appearance
- color
- consistency
- cleanliness.

4.3 PRODUCTIVITY

This is probably the most appealing prospect for the owners of a production process. The prospect of producing 3% more, without increasing unit costs, will have the owners licking their lips (providing they have a competent director of sales and an unsaturated market).

This sounds too good to be true, yet many companies take in raw materials, then spend time to refine them and combine them, only to send an unnecessary proportion of intermediate and even final products to waste. Not only does effluent treatment cost money, you have also lost potential sales. Simply put, your process is more successful if your products are going to your customers and not costing you money to treat as waste.[1]

Productivity losses can also come from plant downtime or plant instability. Often a process can be made to break the shift production record by turning

---

[1] I have seen companies lose 5% of their product to waste, as a result of an old-fashioned filtration and washing process.

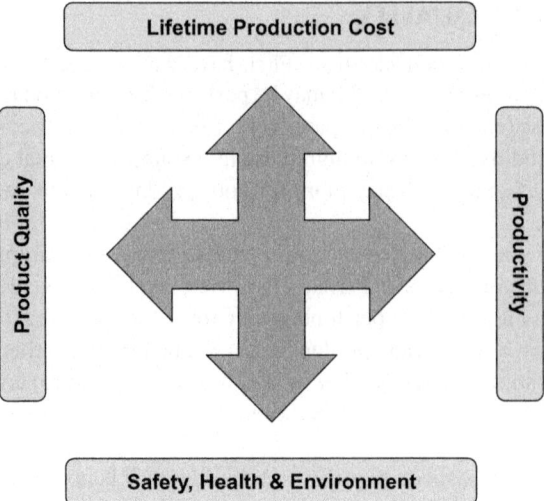

FIGURE 4.1 The four aspects of processing competitiveness.

up all of the dials, only for the process plant to break or the process to become unstable during the next shift. The result can be a poor weekly production total and a large repair bill.[2]

4.4 SAFETY, HEALTH AND THE ENVIRONMENT

Many companies state that SHE (or HSE) are their main priorities: it is often right there on the front page of the company website. Many companies will indeed have a board member responsible solely for SHE, but I cannot think of too many who move from there to become the president or chief executive: it's more likely to be the finance director, for example, who will make that final step up the organization. This is not to say that companies will actively choose profit over a good safety record and therefore put people in more danger, but everyone would rather report excellent financial results accompanied by good SHE data than the other way around: "our profits dropped 5%, but our slippage rate dropped 10%" will not satisfy many shareholders.

Nevertheless, while poor SHE performance can affect your production costs–fines from environmental or safety authorities can be significant, although are often merely a token amount – the main damage is likely to be to your reputation rather than your short-term profit. Recent well-publicized environmental accidents have shown that consumers will actively look for cleaner alternatives. Some people are even prepared to pay slightly more for a product that they feel is less harmful towards the environment, or that is produced

---

[2] I have also seen this happen.

under better conditions for workers. Reputation management is becoming an increasingly widespread idea within the process industries.

Of course, the ultimate sanction for poor SHE performance is that your license to operate can be withheld, having a very significant effect on your productivity, while you attempt to comply and regain the license to operate.

In any case, I have yet to find the plant manager who enjoys appearing on local or national television to explain why the local river or estuary has turned white.

4.5 SUMMARY

The competitiveness of all filtration processes can be expressed in the four dimensions shown in Figure 4.1. This framework can act as a tool to see where you fit relative to your competitors: it is crucial to know this.

This is the first central idea in this book. The following chapter will look at the second central idea: how filtration outcomes link to this competitiveness.

Chapter 5

The Outcomes of Filtration Processes

Chapter Outline
5.1 Filter Cake Outcomes 57
  5.1.1 Cake Moisture 57
    5.1.1.1 Cost of Drying 57
    5.1.1.2 Dryer Throughput 58
    5.1.1.3 Transportation Costs 59
    5.1.1.4 Transportation Moisture Limit 60
    5.1.1.5 Landfill 60
    5.1.1.6 Binding/Pelleting Moisture 60
    5.1.1.7 Conveying 60
    5.1.1.8 Dusting 61
  5.1.2 Cake Washing 61
  5.1.3 Particle Breakage 61
5.2 Filtrate Outcomes 61
  5.2.1 Filtrate Clarity and Volume 62
    5.2.1.1 Product Losses 62
    5.2.1.2 Precipitation or Electroplating 62
5.3 Slurry Outcomes 63
5.4 Filtration Costs 63
  5.4.1 Power Costs 63
  5.4.2 Utilities: Air and Water 63
  5.4.3 Consumables 64
  5.4.4 Maintenance Costs 64
  5.4.5 Operator Costs 64
5.5 Examples of Filtration as a Part of a Process 64
  5.5.1 Mineral Concentrate: Simple Dewatering 64
    5.5.1.1 Outline Description of the Process 64
    5.5.1.2 Importance of Filtration in this Process 66
  5.5.2 Alumina 67
    5.5.2.1 Residue Separation 69
    5.5.2.2 Precipitation Seed Filtration 70
    5.5.2.3 Hydrate Filtration and Washing 71
  5.5.3 Starch Washing and Dewatering 71
5.6 Summary 72

This chapter will link the outcomes of filtration processes to the overall success of the processes in which they operate (within the dimensions of competitiveness framed in Chapter 4). In this way, decisions about filtration processes, during either process development, equipment selection/plant design or ongoing process optimization, will be better informed, more focused and more useful.

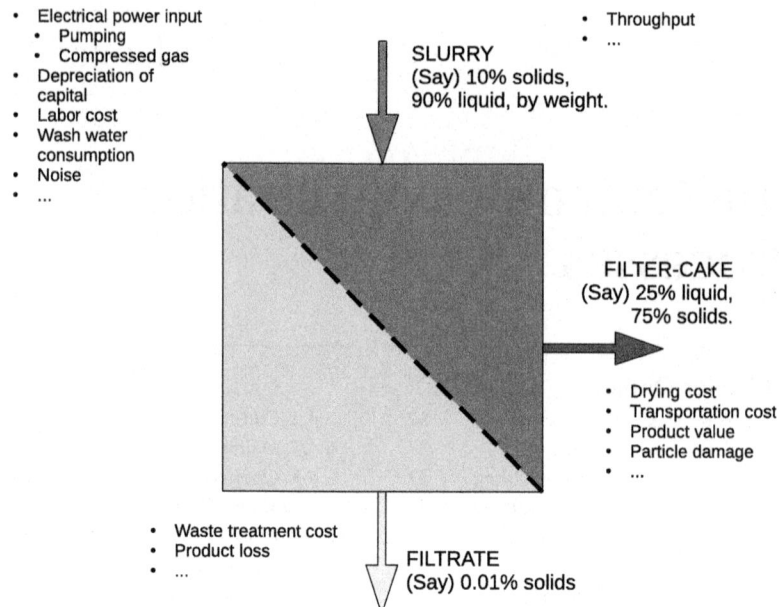

FIGURE 5.1 Detailed description of the outcome of a filtration process.

Some calculation sheets are available for download and these can be used to help quantify the impact of filtration outcomes on the wider process. The outcomes of a filtration process can be categorized as follows:

- Filter cake outcomes, for example:
  - cake moisture
  - wash result
  - particle breakage
- Filtrate outcomes, for example:
  - solids content
  - volume
- Slurry outcomes: simply the capacity or throughput of the filtration process
- Cost of filtration and other consequences, for example:
  - motor power: compressors, vacuum pumps, slurry pumps
  - capital depreciation
  - spares consumption
  - wash liquid consumption
  - noise, dust, etc.

Figure 5.1 shows these outcomes against a schematic of the filtration process itself together with some of the consequences for the overall process.

The following sections will explore the effect that these outcomes may have on the overall success of a process. Where possible, the benefits will be quantified.

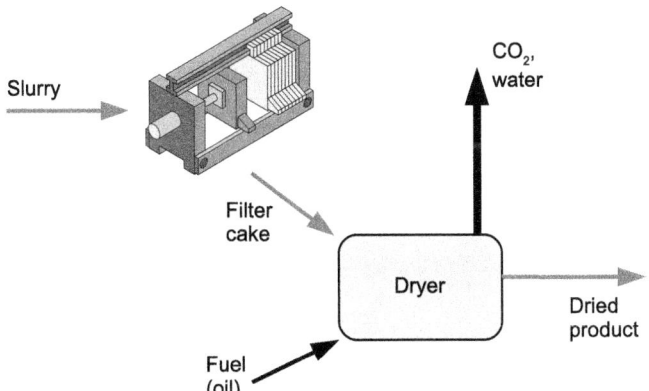

FIGURE 5.2 Fuel consumption related to cake moisture.

5.1 FILTER CAKE OUTCOMES

5.1.1 Cake Moisture

Cake moisture is usually expressed as a weight fraction.[1] The effective cake moisture at various stages of a filtration process, or at various locations on a continuous filter, may be of interest, but it is the moisture at cake discharge that really matters.

Various automatic balances with built-in heaters can be used to measure the moisture of a small sample of filter cake, but the most tried-and-tested method is to weigh the cake before and after drying in an oven.

5.1.1.1 Cost of Drying

If the cake need to be dried after filtration, then the cost of thermal evaporation will be a major factor in the cost of operation. Removing the last amount of moisture can often significantly outweigh the cost of removing all of the liquid up to that point.

As an example, let's look at a process in which 10 tonnes of solid material is filtered per hour, upstream of a thermal dryer. The current filter cake moisture (water) is 20% wt/wt and the moisture required after the dryer is 2% wt/wt (Figure 5.2). A laboratory trial has shown that a modification to the process (i.e. changing one of the six success factors that are discussed throughout Part III, perhaps changing the filter operation by modifying the air-drying pressure) will reduce the cake moisture to 15% wt/wt.

It is straight forward to calculate that the amount of water to be evaporated would be reduced by 0.70 tonnes/hour by this change in cake moisture. Now, if we assume that approximately 120 kg heating oil (or equivalent) is needed to evaporate 1 tonne of water, then the amount of oil (or equivalent) saved would be

---

[1] The weight of liquid in the cake divided by the total weight of the cake (solids plus liquids).

Fuel and Carbon dioxide calculator
www.solid-liquid-filtration.com

10	tonnes/h	Amount cake per hour (dry solids)
8000	h	Operating hours per year
2	% wt/ wt	Required moisture after dryer
20	% wt/ wt	Current cake moisture
15	% wt/ wt	Target cake moisture
2.20	tonnes/ h	Current water to be evaporated
1.49	tonnes/ h	Target water to be evaporated
120	kg	kg oil per m³ water to be evaporated.
3.2	kg/ kg	kg CO_2 per kg oil burnt
250	£/ tonne	price of oil
2107	tonnes/ year	Current annual oil consumption
1434	tonnes/ year	Target annual oil consumption
673	tonnes/ year	Saving in oil consumption
£526,829	£/ year	Current annual fuel cost
£358,621	£/ year	Target annual fuel cost
£168,209	£/ year	Saving in fuel cost
6743	tonnes/ year	Current annual CO_2 produced
4590	tonnes/ year	Target annual CO_2 produced
2153	tonnes/ year	Reduction in CO_2

FIGURE 5.3 A drying cost calculation in a spreadsheet.[2]

0.70 × 120 = 84 kg per hour.[3] Since 1 kg of oil produces around 3.2 kg of carbon dioxide when burnt, 270 kg less will be produced per hour at this lower moisture.

Figure 5.3 shows this calculation in a spreadsheet. If the cost of the fuel used (in this case heating oil) and the number of hours operated per year are known, this sheet gives the annual fuel cost and carbon dioxide production and can be used to explore the cost/benefit of making the change to the slurry that gives the reduction in cake moisture.

This outcome is very significant in terms of production cost and can be weighed against the cost of achieving the lower cake moisture.

5.1.1.2 Dryer Throughput

If the drying process is a bottleneck, and is limited by the amount of water that it can evaporate, then reducing the amount of moisture in the cake will give an increase in solids throughput.

---

[2] See http://www.solid-liquid-filtration.com/
[3] You may have a better fuel consumption figure for your drying system, or can ask the equipment supplier for guidance.

Drying – estimate throughput

Current

Capacity (dry solids)	10	tpa
Filter cake moisture	20.0%	w/w
Moisture after dryer	2.0%	w/w
Wet cake	12.5	tpa
Moisture evaporated	2.3	tpa

Scenario

Cake moisture	15.0%	w/w
dry solids	14.7	tpa
wet cake	17.3	tpa
Moisture evaporated	2.3	tpa
Increase	47.1%	
Increase	4.7	tpa

FIGURE 5.4 Dryer capacity cost calculation in a spreadsheet (downloadable from www.solid-liquid-filtration.com).

As an example, if we take the same scenario above, but in this case assume that the dryer throughput was limited (by the amount of water that can be evaporated) to 10 tonnes/hour at 20% wt/wt moisture, then reducing the cake moisture to 15% wt/wt would give a possible capacity of 14.7 tonnes/hour (dry basis). So, provided the dryer can handle this extra weight (the total weight of cake entering the dryer would increase from 12.5 to 17.3 tonnes/hour, although in both cases 2.3 tonnes of water is evaporated per hour), then this outcome would be viewed favorably in terms of productivity (Figure 5.4).

5.1.1.3 Transportation Costs

Transport does not discriminate over its contents (unless the contents are hazardous). It costs the same (in weight terms) to transport a tonne of water as a tonne of mineral ore concentrate. (In fact, it might cost more to transport the water because of its greater volume.)

In Chapter 1, a simple calculation was given to work out the reduction in number of trucks that could come from reducing the cake moisture:

Say we have a production process producing 10 tonnes per hour (dry basis) of a mineral product, and the wet filter cake is transported 100 km to a customer in 40 tonne trucks. At 25% wt/wt cake moisture, approximately 2920 truck journeys are needed, while at 15% wt/wt, this number would reduce to approximately 2577 journeys, or 343 fewer. The ratio is given by:

$$\frac{1-25\%}{1-15\%} = 0.88$$

Once again, there is a spreadsheet for download. If the current cake moisture, scenario (or possible) cake moisture and current transportation costs are input, the sheet returns the possible savings in production cost.

5.1.1.4 Transportation Moisture Limit

Many mineral processing industries define a transportation moisture limit (TML). Above this moisture, the cake could reslurry under the vibrations and periodic motions of trucking or shipping. In extreme cases, this could cause the load to slosh about and make the truck or vessel unstable. This would be a significant safety, health and environment (SHE) issue, not to mention production cost.

5.1.1.5 Landfill

Governments around the world are introducing landfill taxes. These are often based on the weight, or volume of material placed in landfill.

So, if the filter cake is a waste destined for a hole in the ground, you will also be paying for the water to go in. As before, a spreadsheet is supplied to estimate the cost impact of reduced cake moisture on landfill charges.

5.1.1.6 Binding/Pelleting Moisture

In some cases, the filter cake may be blended with other components and formed into pellets, before transportation to (say) a smelter. In this case there may be an optimum moisture at which the ingredients bind together to form robust pellets. If the cake moisture wanders too high or low then the pellets could be either too hard or too crumbly.[4]

It is more difficult to quantify a benefit accruing from having the correct cake moisture for binding, but having a filtration process that reliably delivers cake within the required moisture level (not too wet, not too dry) will provide a major benefit in terms of product quality, which in turn affects productivity because of the amount of unusable product.

5.1.1.7 Conveying

Filter cake can be some of the most difficult material to convey, not least because it can potentially reslurry or sometimes "sweat" moisture. If this happens then a situation can arise where a belt conveyor is sliding underneath a stationary block of filter cake, and all the time new cake is added to the pile.

---

[4] In one extreme case (on a steel plant, the process was making briquettes of material recovered from a scrubber to be fed back into the steel blast furnace), if the moisture of the cake was too low, the cake oxidized exothermically; if this exothermic reaction was not inhibited enough by moisture then the cake "burnt".

Also, there can be a great deal of difference between a continuous filtration process that delivers a steady feed of cake onto a conveyor and a batch process that delivers several tonnes, in one go, say every twenty minutes.

The nature of filter cake, its moisture, etc., can have an impact on the capital and operational cost of conveying (a production cost issue) or the reliability of a conveying operation, which can, in turn, affect productivity.

5.1.1.8 Dusting

As before, if a filter cake is too dry then this can give rise to dusting, an issue for the SHE performance of a process. The method of cake discharge can be a factor here: in some tower presses (Section 7.3.2) the cake may fall for up to seven or eight meters before hitting a conveyor.

If the cake is discharged as a batch, the falling cake will displace air in the chute and this can produce a dusty updraught.

5.1.2 Cake Washing

Cake washing results can affect a process in a number of different ways. If the product, or another raw material, is trapped in a filter cake that goes to landfill then this can become a major production cost or productivity issue. One example of this is discussed later in this chapter (see the discussion on red-mud filtration in Section 5.5.2).

It should be relatively simple to calculate product losses, as well as potential savings or increases in productivity that could come from an enhanced cake washing process.

It is not only the best achievable cake washing result that is important, but also the consistency or product quality. As discussed in Chapter 4, it is not good enough to produce a food that is the most delicious and wholesome that money can buy for 99% of the time, but for the remaining 1% may cause terrible stomach upsets.

5.1.3 Particle Breakage

Breakage of particles can have a major impact on a product's critical quality parameters. For example, if the cake is an active pharmaceutical ingredient, destined to be absorbed in the human intestine, then the rate of absorption will be changed if the particles in the cake have been broken into pieces: they would have a much higher net surface area and would dissolve too quickly.

Particle breakage can also affect dusting.

5.2 FILTRATE OUTCOMES

There are two main outcomes to be considered in terms of filtrate: clarity, or solids content, and volumetric flow. The two factors go hand in hand and will be considered together in most of the following examples. (There may be other, more subtle outcomes to consider, such as the temperature of the filtrate if it is going on to a precipitation process, but these are too specific for this section.)

5.2.1 Filtrate Clarity and Volume

The clarity of filtrate, or the solids content, can be measured relatively easily and has a profound impact on competitiveness. Some examples are given below.

5.2.1.1 Product Losses

If the solid phase is the product and if the filtrate goes to waste treatment then any solids carried away will be lost. For example, if a process is currently using 2 tonnes of wash water per hour and the filtrate contains 0.25% wt/wt solids, then adopting a technology that uses less than 1 tonne of water per hour and produces filtrate with a solids content of 0.01% wt/wt makes a huge difference in terms of productivity (the solid product goes to the customer, not the waste plant) and production cost (the waste treatment costs more than halve, as does the amount of wash water consumed).[5] In this case, the losses to waste reduce from about 43 tonnes per year to less than 1 tonne per year.

Capturing solids in the filtrate from one filter using a polishing filter can often be justified simply in terms of the value of the product recovered and not sent to waste. This is before the reduction in waste treatment costs is considered.

5.2.1.2 Precipitation or Electroplating

Many processes produce solid products by precipitation or crystallization from a liquid. For example, if a white product is precipitated from a solution that contains suspended material, then the product may be spoiled – sugar that is full of black specks is not desirable to the tea drinker.

The bulk of the world's copper is produced by electroplating copper from solution onto stainless steel cathodes in vast baths of copper electrolyte. Any solid contamination in these baths will reduce the quality of the copper produced. In addition, any solid contamination may settle onto a cathode and a nodule of copper will grow over it (like an oyster producing a pearl). This nodule gives a short-circuiting bridge for the current flowing between the anode and cathode, reducing the energy efficiency of the process. Filtration of the liquid in these baths (arranged as a side loop) reduces the amount of solid material in the bath, improving both product quality and energy efficiency/production cost.

Filtration is a vital pretreatment for the brine in the chloralkali industry (in which products such as chlorine and sodium hydroxide are produced by electrolysis of brine). The presence of solid matter in filtrate can foul the permeable membranes

---

[5] This was a tremendously successful argument from the days when I was involved in sales and marketing of filtration equipment: the application was food starch washing and dewatering and high solids in the waste water system also produced a terrible smell.

in the electrolysis cells. Almost absolute filtrate clarity is necessary to prolong the lifetime and energy efficiency of the membrane cells.

Finally, if the filtrate and wash filtrate are a pregnant liquor (containing a dissolved product that will be precipitated) then the stronger the concentration the better. So, a washing process that minimizes wash liquid consumption while maximizing recovery will be required.

5.3 SLURRY OUTCOMES

The main consideration here is simply the capacity, i.e. the throughput in terms of slurry processed. If the central stream of the process passes through the filtration process, then any disruption in capacity will clearly affect productivity.

A key consideration here is the availability of the filtration process. Time lost to planned or unplanned shutdowns should be included.

The operating mode of the filtration equipment is also crucial. A continuous process may need substantially smaller slurry and cake handling systems.

5.4 FILTRATION COSTS

Most of the costs outlined below are self-explanatory. In the process of choosing a filtration system for a process, the cost of filtration, probably expressed per tonne, is important, but it is crucial to investigate the effect of any filtration process on the wider process and not simply to make comparisons based upon the capital and operating costs of each proposed filtration solution.

5.4.1 Power Costs

This is simply the electricity bill for the filter itself as well as all of the auxiliaries that surround the filter, including slurry pumps, conveyors and other drive motors.

5.4.2 Utilities: Air and Water

The price of compressing air is significant and, in most cases, all of this pressurized air will ultimately be vented back to the atmosphere at the back of the plant. It is important that this air goes to good use. In some cases the air-blowing stage of a filter may have crept up over the years, and may be longer than the point at which it no longer makes much difference to the final moisture.

It is well worth checking the consumption of compressed air at frequent intervals, as this may offer an immediate opportunity to make a saving in filtration costs.

Equally, it is worth checking that the quantity of wash water is appropriate (and that the wash water consumption places you on the long asymptotes of the wash curve). It is not simply the cost of consumption of wash water to consider, but the impact that the higher volumes of wash filtrate or greater level of product dilution will have on the process steps that follow.

5.4.3 Consumables

The cost of all spare parts and consumables, including cloths, seals and replacement mechanical parts, should be considered. Chapter 10 contains a discussion on how the cost of consumables may be affected by operational choices.

5.4.4 Maintenance Costs

These include the cost of all labor required for routine and unexpected maintenance of the filtration equipment. In equipment selection and plant design, maintenance issues will require close discussion with the equipment vendor and other users.

5.4.5 Operator Costs

Some filtration processes operate without any intervention at all from operators, others may require supervision (e.g. to ensure that all of the cake discharges at the end of a batch) while others may need an operator to discharge the filter cakes (as in Figure 2.3).

This factor should be included in the assessment of any filtration process. There is also a point to be made in terms of SHE, especially if the materials being filtered are toxic or unpleasant.

5.5 EXAMPLES OF FILTRATION AS A PART OF A PROCESS

5.5.1 Mineral Concentrate: Simple Dewatering

The global mining industry extracts hundreds of millions of tonnes from the Earth's crust as the first step towards producing useful minerals and metals.

A mined copper ore can contain less than 1% copper by weight, combined with the balance of unwanted rock.[6] It is uneconomical to extract copper directly from this ore, or to transport all of it to a refinery, so the copper fraction of the ore is usually liberated from the ore at the mining location. The concentrated fraction is then transported to another location for further refining to the metal and the residue left at the mine site.

5.5.1.1 Outline Description of the Process

Figure 5.5 outlines the main steps in this process:

- Crushing/grinding: The rock is crushed and ground to a size such that each individual particle of rock either contains copper or does not (Figure 5.6).
- Flotation: The crushed ore is suspended in water and pumped to flotation columns. These exploit the fact that the copper-bearing grains tend to attract fine bubbles of air blown through the columns, while the non-copper-bearing

---

[6] Known as gangue.

Chapter | 5 The Outcomes of Filtration Processes   65

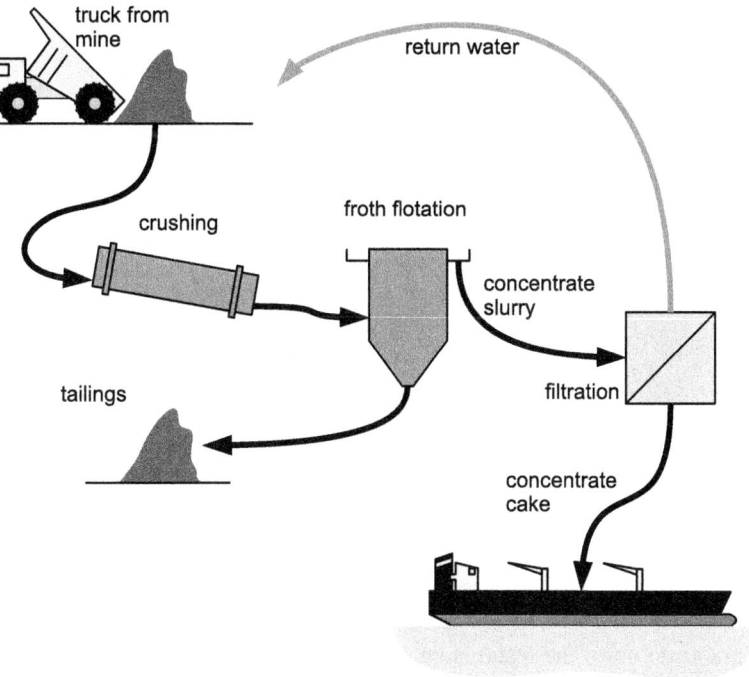

FIGURE 5.5 Simplified flowsheet of a mining process.

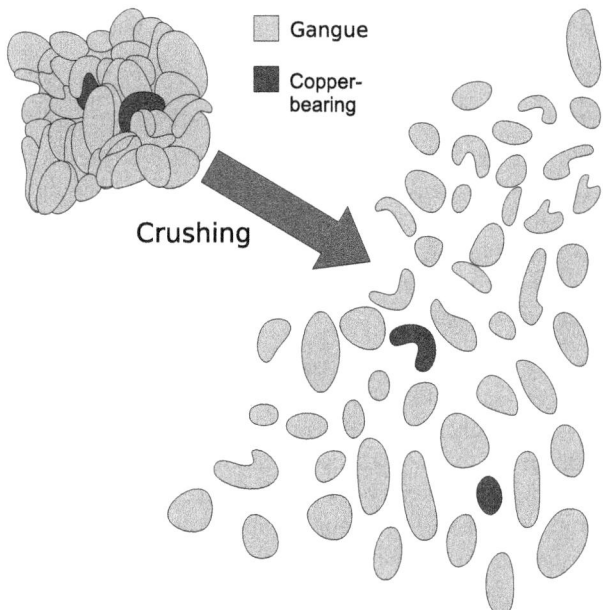

FIGURE 5.6 A piece of copper ore, before and after grinding to liberate the copper-bearing part.

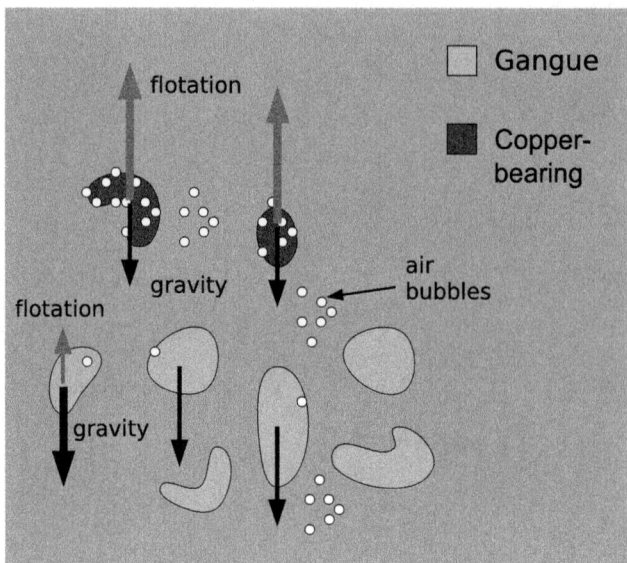

FIGURE 5.7 The principle of flotation.

grains do not.[7] The result is an overflowing froth of copper mineral and an underflow of gangue (Figure 5.7).

- Filtration: Flotation has served the purpose of enriching the copper mineral, but this is now contained in a slurry, an unsuitable form for further processing or transportation. Filtration is used to remove as much water as possible. Thermal drying may also be needed.

5.5.1.2 Importance of Filtration in this Process

If filtration did not exist, humans could still produce copper. It would just be astronomically more expensive and environmentally damaging to do so. An alternative to filtration would be to use huge gravity thickeners to remove as much water as possible then wait (luckily many of the world's copper mines are in dry, sunny parts of the world).

Some other considerations:

- If the moisture content is above a certain level (the TML), there is a large risk that the concentrate could reslurry during transport. If a ship, designed to carry a static load, suddenly holds a large massive body of dense slurry, it is possible that the stability of the ship will be jeopardized.
- Water balance: It is not very useful to transport water from, say, Chile or Australia to Europe. In fact, many mining areas are water depleted, so you would need to transport water to the site, only to take it away again.

---

[7] This effect is enhanced with certain flotation-aiding chemicals.

5.5.2 Alumina

Aluminum is an extraordinarily versatile and economically important material.[8] Although it is abundant within the Earth's crust, the planet does not want to give up this silver-colored, light, strong and corrosion-resistant metal easily. While its ore (bauxite) is relatively easily extracted, usually with surface mining, the process of refining the metal requires a great deal of effort. It is estimated that aluminum production accounts for around 1% of anthropogenic greenhouse gas emissions. However, these are somewhat offset by the reduction in greenhouse gases emitted from cars, lorries and trains made from aluminum rather than steel.

The normal production route taken is to refine the bauxite to aluminum oxide, or alumina (Al_2O_3), a white, crystalline powder, and then to smelt this powder to aluminum metal. Whereas aluminum smelters tend to be located close to where the aluminum is needed, alumina plants are usually closer to the source of bauxite, or en route to the aluminum smelter. (Large quantities of fuel and sodium hydroxide are also required, and their location and transportation costs are also factors in deciding the location of alumina plants.)

The production of alumina from bauxite is a hugely significant industry in its own right but it is also one of the most significant global users of filtration and separation technology. A typical alumina plant may use approximately 500–1000 m^2 of filtration area per million tonnes of production (as well as many hundreds of square meters of gravity thickener/clarifier area). Since total production of alumina is around 80 million tonnes per year, there are many tens of thousands of square meters of filtration operating on alumina plants at the precise moment that you are reading this.

Production of alumina approximately doubled between 1990 and 2010 (see Figure 2.4). A significant proportion of this increase came from squeezing more capacity out of existing plants or adding additional streams to these plants.

Almost all alumina plants in the world use the Bayer process, patented over 120 years ago,[9] to refine bauxite to alumina. In this process, a large volume of caustic liquor recirculates continuously around the plant (Figure 5.8). Bauxite is fed into the caustic stream and, after a number of processes, alumina is taken out of the stream. The main process steps are:

1. Dissolution of the aluminum-bearing minerals in a caustic liquor at high temperature and pressure.
2. Removal of the solid residue (the non-aluminum-bearing part of the bauxite, usually a mixture of iron-rich minerals).
3. Precipitation of pure alumina hydroxide [$Al(OH)_3$], under conditions of controlled cooling.
4. Calcination of the alumina hydroxide to remove the water of crystallization so that it is ready for the aluminum smelter.

---

[8] This section is adapted from Sparks (2010).
[9] See, for example, US Patent 515,895, Process of making alumina, Karl Bayer.

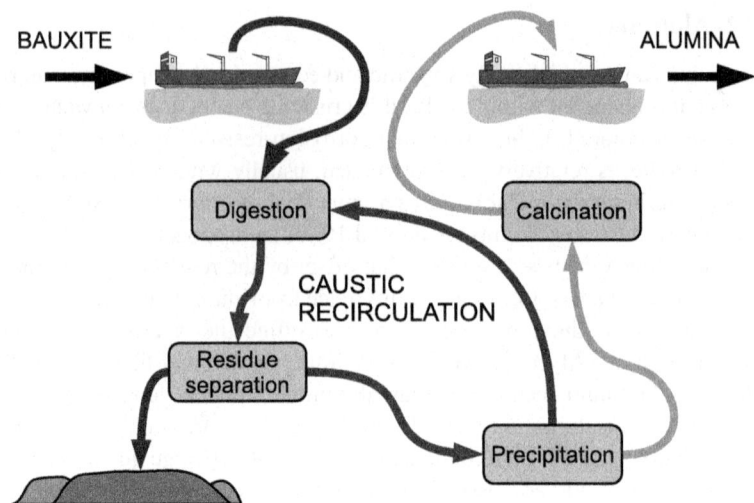

FIGURE 5.8 Simplified alumina flowsheet.

So, here we have a process that uses a concentrated caustic liquor at high temperature and which produces an extremely abrasive product – seemingly not an ideal location for filtration equipment. However, filtration equipment is operating successfully on alumina plants in some of the most harsh conditions found anywhere in the process industry.

A further interesting aspect of this process, as we shall see, is that it depends upon each of the basic motivations for solid–liquid filtration types:

- separating valuable liquids from less valuable solids (red-mud separation)
- separating valuable solids from less valuable liquids (final product filtration)
- separating valuable solids from valuable liquids (seed filtration).

As with all production processes, production cost, product quality, productivity and SHE ultimately define its success. For an alumina producer, some of the more important considerations are:

- Cost:
  - efficient use of heat, caustic, capital employed and water.
- Product quality:
  - removal of non-aluminum compounds (e.g. the presence of iron or titanium compounds adversely affects aluminum production)
  - consistent alumina particle size distribution (particles that are too fine will cause dusting; too coarse and they will disrupt the electrolytic smelting process).
- Productivity:
  - minimizing product losses to the environment and waste
  - operating continuously (24/7) for as many days per year as possible.

- SHE:
  - managing the risks associated with caustic materials, and very high temperature and pressure
  - managing the large quantity of residue produced.

The four most significant solid–liquid filtration processes are:[10]

- residue separation – red-mud filtration
- residue separation – liquor polishing
- precipitation seed filtration
- product washing and filtration.

5.5.2.1 Residue Separation

The main aims in this step of the process are to produce a clarified solution that can be delivered to the precipitation process and to produce a residual mud, containing a minimal amount of process liquor, that can be safely stored, or disposed of, by stacking.

This clarification is normally achieved through gravity settling clarifiers, with a final polishing filtration step for the overflow from the final clarifier and washing and filtration for the underflow (the red mud). If you look for satellite images of alumina plants on the internet, the red-mud disposal areas are unmistakable (e.g. look south-east of St Petersburg in Russia, or around Perth in Western Australia).

Given the composition of bauxite, for every tonne of alumina produced there is approximately another tonne of red mud. Naturally, there is a certain advantage to refining alumina close to the bauxite mining activity, since the red mud can be used to backfill the mine. However, it is important that the quantity of caustic material in the mud is minimized and that the disposal area is sealed to prevent leaching of caustic soda into the surrounding soil.

Filtration in the Process – Red Mud

In most circumstances, plants use vacuum filters, usually drum filters, for the final washing and deliquoring of red mud before it is sent over the fence to a waste area. In this process, the solids are waste and the liquor is valuable. The reasons for using drum filters are their reliability and ability to discharge thin, sticky cakes. The feed to the filters at this stage may have a solids content of approximately 40% wt/wt and the liquor contains a quantity of caustic soda and dissolved aluminum.

Future Trends in Red-Mud Filtration

There have been numerous attempts to find uses for red mud. (Karl Bayer's original patent suggests that the red-mud residue could be used in the production of iron.) However, so far, there have been very few successes in this

---

[10] There are others, as well as gas/solid filtration duties.

endeavor. Some current initiatives are pursuing the idea of exploiting red mud, for example as a construction material (see www.redmud.org). Some of these applications would depend upon complete removal of caustic and other liquors from the mud.

Filtration in the Process – Liquor Polishing

The accumulated overflows from gravity settling may still contain a small amount of fine suspended solid matter. If these particles were allowed to pass into precipitation, they would contaminate the product with, for example, iron and titanium compounds. This would then affect the properties of the alumina and, ultimately, affect the aluminum smelters. For this reason, the liquor is passed through polishing filters to remove this matter.

Of all the filtration applications in alumina, security filtration appears to be getting the most attention from equipment suppliers at present. The industry is constantly looking for automatic, reliable and self-cleaning equipment for this application. Many filters installed in this application require a great deal of attention, this is mainly because of scaling (precipitation onto the internal surfaces and media within the filters).

5.5.2.2 Precipitation Seed Filtration

During his original development work, Bayer mixed a dose of aluminum hydroxide crystals into the sodium aluminate solution to provide a seed for precipitation. This technique is still used today, on a scale several orders of magnitude greater than the glassware used in St Petersburg during the 1880s.

In the modern interpretation of the Bayer process, vast tanks are used to give the pregnant liquor a long residence time in the presence of seed particles and under carefully controlled cooling so that the solution can give up its product to the solid phase. In another largely exploited precipitation process, the manufacture of precipitated calcium carbonate, the time required for the crystals to form is almost zero. In the case of alumina, several tens of hours can be required to produce the carefully controlled particle size distribution required by the later process of making metallic aluminum through smelting.

Once the liquor has given up all of the product that it can (and therefore become spent liquor), it is separated from the solid product (usually in large settling tanks) and returned to the beginning of the Bayer circuit, where it is reheated and reconcentrated, ready to meet fresh incoming bauxite. The solid particles at the end of the precipitation chain are classified, with those meeting the required product size being diverted to the calcination step and those undersize being returned to the top of the precipitation chain to act as a seed for precipitation.

There is an advantage to stripping out the spent liquor from the slurry of seed that passes back up the precipitation chain in order to maintain a high concentration (and therefore more precipitation potential). This gives the plant a productivity boost.

Filtration in the Process – Seed Filtration

The seed slurry in this application filters under vacuum in a few seconds to produce a very substantial cake, in some cases more than 50 mm thick. In most cases, vacuum filtration is used for this application, occasionally drum filters or, more typically, disc filters.

This is not a challenging filtration process in terms of the difficulty of the separation itself, but rather in terms of handling very large throughputs of heavy, abrasive cake in a highly caustic mother liquor. A further challenge is handling very large quantities of filtrate; in other words, the filtrate system on the filter itself and throughout the installation must be well designed.

5.5.2.3 Hydrate Filtration and Washing

The graded and classified solid product from precipitation must be washed to remove the process liquor which would contaminate the final product and interfere with the smelting process. The aims here are to produce a filter cake that is dry enough to be fed to the calciner and to wash the cake free of process liquor.

Filtration in the Process

Most commonly, pan filters are used for this application. They are used because of their relative compactness and cake-washing capabilities, together with their inherent reliability and ability to handle hot wash liquids.

Alumina refining is one of the most significant global filtration applications, in terms of the number of filters operating. As with copper mining, it would be possible to make alumina without filtration, but it would be far more expensive and environmentally damaging, and it is unlikely that the product would be consistently of high enough quality.

5.5.3 Starch Washing and Dewatering

All of the energy for life on Earth originated in the sun and is converted to food by plants, which store their energy as a complex glucose polymer, starch. Starch is used in many food stuffs and flours, and as a filler in some plastics or paper/board products. The crops grown commercially for their starch (in approximate order of slurry filterability, from relatively easy to relatively hard) are potato, tapioca, corn, rice and wheat.

Where the properties of the starch particle need to be modified, for example so that they bind to the fibers in paper or blend more easily into a baby food, chemical or heat treatment may be used. Any such chemicals must be removed before the starch can be used, and a filtration washing and dewatering process is used at this step.

Some important drivers for the competitiveness of a starch refining process are:

- Cost:
  - efficient use of heat, chemicals, capital employed and water.
- Product quality:
  - removal of modifying agents that could affect the taste, performance or even safety of the products made using the starch
  - if the starch product meets the requirements to be classified as food grade, then the price can increase dramatically.
- Productivity:
  - minimizing product losses to waste disposal
  - operating 24/7 for as many days per year as possible.
- SHE:
  - reducing the carbon footprint of the plant
  - minimizing the smell (rotting starch is unpleasant).

5.6 SUMMARY

The outcomes of filtration processes, in terms of:

- cake outcomes
- filtrate/wash filtrate outcomes
- slurry throughput
- cost of filtration

can affect the overall success of a production process. However, throughout industry, in many cases the filtration outcomes are not considered fully in the light of their effect on the success of the overall process.

This chapter discussed this in some detail, without really looking at how these outcomes might be achieved. That is what Part III aims to do.

Part III

Filtration Process Success Factors

Chapter 6　Slurry Filterability　　　75
Chapter 7　Filter Design　　　81
Chapter 8　Filter Installation　　　125
Chapter 9　Filter Cloth　　　131
Chapter 10　Filter Maintenance　　　147
Chapter 11　Filter Operation　　　153

The following chapters will explore the impact that the success factors identified in Chapter 1 have on the outcomes of filtration processes.

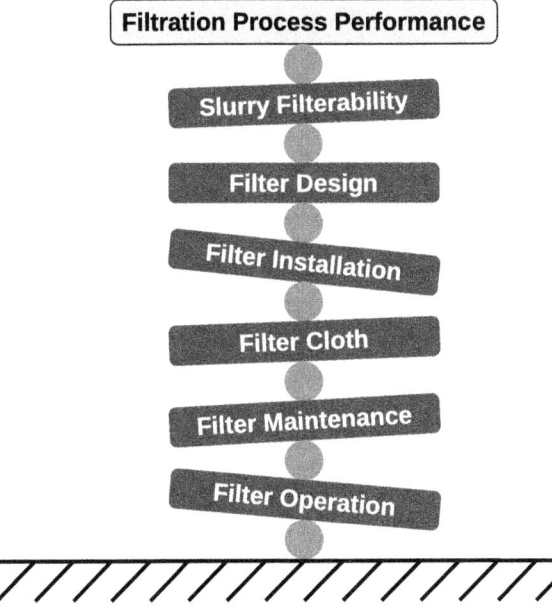

Chapter 6

Slurry Filterability

Chapter Outline

6.1 The Nature of the Slurry to be Filtered	76	6.2.3.1 Chemical Additives	78
6.2 Pretreatment of the Slurry	76	6.2.3.2 Body Feed	79
6.2.1 Density	77	**6.3 Slurry Handling**	**79**
6.2.2 Temperature	77	6.3.1 Pumping	79
6.2.2.1 Viscosity	78	6.3.2 Storage/Suspension	79
6.2.2.2 Particle Rigidity	78	6.3.3 Flow Control: Valves	80
6.2.3 Additives	78	**6.4 Summary**	**80**

It may seem quite trivial and self-fulfilling to include slurry filterability as a factor in the success of a filtration process. However, there are ways in which the success of a filtration process can be improved by looking at the material that is delivered to it; it is wrong to assume that the slurry is "just the way that it is".

Changing the process upstream can deliver significant improvements and, in a perfect world (from a filtration plant operator's point of view), all processes could be "engineered for filtration", or designed to produce a slurry that filters optimally. Unfortunately, during the design of production process, filtration is not usually among the primary considerations. It is not unusual to see a process design in which it is established that filtration is possible, using a laboratory Büchner, and the matter left there.[1] Furthermore, it may simply not be possible, no matter how much we would like it to be, for filtration to be the primary concern. I witnessed an argument over this once, where the product (a mineral paper filler) was milled to reduce the particle size before filtration; the filtration expert involved could not understand why it could not be done afterwards – it could not because the performance of the end product would not be achieved that way.

---

[1] Filtration has not been called a "Cinderella technology" (Wakeman and Tarleton, 2005b) for nothing.

It also depends where you are in the life cycle of the product: process development, equipment selection/plant design or process optimization. The following sections explore the issues and suggest where and how changes could be made.

Chapter 3 discusses the microscopic physical phenomena of a filtration process with the idea that an appreciation of these phenomena can provide the path to improvement. The nature of a slurry affects the performance of a filtration process, but does not necessarily make it possible to predict this performance quantitatively. Progress can be made by using thought experiments to explain what you see at the macroscale and using tests to examine these explanations. However, you should also be prepared for the unexpected.

6.1 THE NATURE OF THE SLURRY TO BE FILTERED

The question here is: "Can we make a fundamental change to the nature of the slurry arriving at this filtration process?" Is it possible to change the particle size distribution by delaying milling until after the filtration step, or is it possible to produce larger particles by modifying the precipitation or crystallization process? Is it even possible to use a different liquid in the slurry? In short, is it possible to re-engineer the process so that that filtration comes further up the list of considerations?

In some cases a product may have a critical quality parameter that is directly related to the filterability of the slurry. Examples might be the light-scattering properties of a pigment or the rate at which a drug dissolves in the gut, both of which are largely determined by particle shape and size, which in turn both strongly affect filtration and washing rates. In this case, options for modifying the particle size and shape will be limited.

If particle size does not really affect the rest of the process, for example if the precipitated or crystallized solids are an intermediate step and will be redissolved later in the process, then it could make sense to review particle size and consider the benefits of larger particles. These could include reduced capital spend (a smaller filter), lower cost of operation, longer spares lifetimes (especially cloths), lower cake moisture, reduced losses to filtrate or better cake washing (in terms of final result, wash liquid consumption, or both).

6.2 PRETREATMENT OF THE SLURRY

If it is not possible to change the nature of the slurry in a significant way, then there are still ways in which the filtration process performance can be improved by slurry pretreatment. There will be some cross-over here, but fundamentally we are not looking to change the particle size distribution or liquid composition.

6.2.1 Density

The density, or solids content, of the slurry to be filtered is one of the most significant factors in filtration outcome, not just in terms of throughput, but also in terms of spare-part lifetime (Section 3.2.2).

The effect of solids content in the slurry, or slurry density, can have a significant effect on the performance of a filtration process (Chapter 3). If the solids content of a slurry is low then the particles transferring into the cake will not be competing for space as much as they would be if the solids content were much higher. Therefore, there will be less bridging and a less permeable, less structured cake will result.

This can have a major effect upon the filtration rate. A fall from, say, 25% solids by weight to 20% solids can have a huge impact on filtration rate (even if it can be difficult to predict precisely how much). A laboratory-scale test could give a rapid indication of this impact.

The lifetime of filter cloths can also be affected significantly by changes in the slurry density. In general, if the filtrate contains a large amount of solids, and if these solids are sharp or abrasive, then the cloth lifetime will suffer if the slurry solids content reduces significantly. On some industrial minerals plants, the filtration process is automatically stopped as soon as the density drops, if say a thickener deviates from its usual performance, because the alternative would be a new set of filter cloths.

Slurry density can be increased by utilizing upstream thickening, using hydrocyclones, sedimenting centrifuges or gravity thickening. It may also be improved using coagulants or flocculants. If a filtration plant is at the limit of its capacity then any one of these measures could be an alternative to installing more filtration capacity.

6.2.2 Temperature

Temperature can be shown to be a significant factor in filtration performance, mainly because of the reduction in liquid viscosity, but also sometimes because of changes in particle characteristics. In any case, it is important to make sure that the temperature is specified when testing, so that the effect of any changes can be investigated.

The thermal efficiency of the process industry is a major concern and great improvements have been made in the past few decades.[2]

Although heat exchange with a slurry can be difficult, it may not be impossible, and retaining more heat within a slurry should be relatively simple.

---

[2] Cefic, the European Chemical Industry Council, maintains a wide variety of database of information on energy efficiency, which can be seen at www.cefic.org/Facts-and-Figures/

6.2.2.1 Viscosity

It should be possible to calculate the increase in the temperature of slurry reporting to a filtration step from lagging the pipeline. It is then very simple to perform some small-scale tests, at both this new, possible temperature and the existing temperature, to determine the change in filtration performance. If this leads to an increase in throughput then this can be weighed up against the cost of lagging the pipeline.

The lower the liquid viscosity, in general, the lower the force on particles dragged onto the filter cake, and the cake formed will tend to be more open, having a higher porosity (void ratio) and permeability. A less viscous mother liquid will also pass more freely (with less power required).

6.2.2.2 Particle Rigidity

The majority of solid particles within the scope of this book are rigid, or crystalline. However, if the particles are gum-like, then there could be advantages to cooling the slurry before filtration, and these advantages can outweigh the penalty of increased viscosity. I have seen one case where a food product could hardly be filtered at all at high temperature (when the gummy particles formed an impenetrable coating to the filter cloth). At a much lower temperature, when the particles became more rigid, the material filtered well, to the point where the process became viable.

In regions with cold winters, for a large part of the year the cooling of a slurry will not cost much more than diverting a pipeline out of the building.

It some circumstances the solid component of a slurry may melt if the temperature increases. An obvious example is the filtration of ice crystals, where an increase in temperature would destroy the solid product. Another example is the fractionation of palm oil: below a certain temperature the stearin fraction of palm oil becomes solid and can be filtered to remove it from the olein (liquid component).

6.2.3 Additives

Chemical or physical additives can transform a process. Many suppliers have mobile laboratory equipment or their own laboratories that they can use to assess the effect of their products on the filterability of slurries.

6.2.3.1 Chemical Additives

Chemical additives can work in a number of different ways to enhance filterability (or washability) by modifying the way that particles interact with each other, whether or not they form tight agglomerates or looser flocculated clumps.

Chemicals may also be added to wash liquids, often in tiny quantities, to modify the surface tension between liquids and particles in the filter cake. These can offer the benefit of reducing final cake moistures while reducing the flow of air through the cake. The penalty is the introduction of a contaminant to the product.

6.2.3.2 Body Feed

The use of body feed was discussed in Section 3.3.5. The significant global pre-coat and body feed industry points to the fact that this can be a successful way to improve the filterability of a slurry. The major downside is that this will contaminate the cake, so it is rarely an option if the solids in the cake are the product.

In certain circumstances, especially if the particle size distribution is bimodal (e.g. there is a distinct coarse fraction), there may be benefit in classifying this slurry (e.g. using a hydrocyclone) and laying down a bed of the coarse material before feeding the finer material. Something similar to this effect can occur in slurries with a very coarse particle fraction, such as alumina hydrate. The rate of settling is so fast that the coarse material can settle onto the cloth before the filtration process has really begun.

6.3 SLURRY HANDLING

The structure of any flocculated particle groups can be disrupted if the slurry is sheared or churned excessively en route to the filtration process. Equally, some fragile particles, for example highly structured precipitated calcium carbonate particles, can be broken with rough treatment. Three important issues upstream are pumping, storage and flow control.

6.3.1 Pumping

The flow through centrifugal pumps, especially when they are operating away from their best efficiency point, can be highly disordered. A great deal of the energy supplied by the motor goes into mixing and shearing the contents of the impeller casing. The rate of feeding to most pressure filters will vary greatly during a filtration cycle, from the beginning of the cycle, when the filter is being filled against almost no resistance, to the end of the cycle, when the cake resistance is at its highest and the flow has dropped significantly. This means that for long periods the slurry is treated harshly in a high-shear turbulent-flow regime.

Advice should be sought from pump suppliers if this is thought to be an issue. Alternative pump types can be considered, for example progressive cavity pumps, hose pumps or diaphragm pumps, although care must be taken to avoid pulsations, or pressure shocks, in the flow to a filter since these can disrupt the structure of a fragile filter cake or cause structural harm to pressure vessels.

6.3.2 Storage/Suspension

Slurries are often stored upstream of the filtration process and normally the slurry is agitated to prevent settling. Care should be taken to ensure that fragile particles (or groups of particles) are not damaged by this suspension. Agitators range from small-diameter, high-speed, axial-flow types (with high shear) to, for example, slower moving anchor agitators.[3]

6.3.3 Flow Control: Valves

A wide variety of flow control valves is used in filtration plants. A partially opened ball valve or butterfly valve will impart high shear to a slurry passing through it, and this should be considered. Of all valve types, pinch valves probably offer the most gentle regime for slurry flow control. These valves comprise a flexible hose with the same bore as the pipeline in which they sit, so when the valve is fully open the slurry does not experience any disruption. Flow is controlled (or stopped) by squeezing the outside of the hose.

Quick particle size distribution tests (on samples taken before and after) should indicate whether these are significant issues on an existing plant. These factors should be considered during process design, and the sensitivity of a slurry to shear can be tested in the laboratory.

6.4 SUMMARY

It can be dangerous to assume too much about the filterability of a slurry. However, in general, the filterability of a slurry can be improved by:

- increasing the temperature of the slurry (or not allowing it to cool).
- increasing the density of the slurry
- increasing particle size – if possible or allowable:
  - using slurry additives
  - using chemicals to modify pH or cause the particles to group together
  - by bulking up the slurry with solid matter.

These can, and should, be tested at small scale.

The key question is how any improvements in the overall process (framed in terms of cost, quality, safety, health and environment, and productivity) compare to any additional costs.

It is hoped that the idea of creating more filterable slurries is a concept that can be adopted more in future, so that the Cinderella technology can go to the ball more often.

---

[3] In one case, I was present at the commissioning of a large and rather expensive filter. Throughout the selection process, the slurry filtered well, producing a dry, crumbly cake (the product, an ingredient for a pesticide) and an extremely clear filtrate. One year later, when the filter was commissioned, the performance, with the same filter cloth and operating conditions, was very poor; in fact, it was difficult to tell the filtrate from the slurry. The main cause was a change in the impeller used in the precipitation tanks, giving a much finer particle size.

Chapter 7

Filter Design

Chapter Outline

7.1 Vacuum Filtration: Continuous 84
  7.1.1 Rotary Vacuum
    Drum Filter 84
    7.1.1.1 Slurry Feeding/
      Filtration 86
    7.1.1.2 Washing 87
    7.1.1.3 Cake Pressing
      (Optional) 87
    7.1.1.4 Air Drying 88
    7.1.1.5 Cake Discharge 88
    7.1.1.6 Cloth Cleaning 89
    7.1.1.7 Installation 89
    7.1.1.8 Maintenance
      Notes 90
    7.1.1.9 Options/
      Alternatives 90
    7.1.1.10 Applicability 90
    7.1.1.11 Precoat Filters 90
  7.1.2 Rotary Vacuum
    Disc Filter 91
    7.1.2.1 Washing 92
    7.1.2.2 Air Drying 92
    7.1.2.3 Cake Discharge 93
    7.1.2.4 Cloth Cleaning 93
    7.1.2.5 Applicability 93
    7.1.2.6 Installation Notes 93
    7.1.2.7 Maintenance
      Notes 93
    7.1.2.8 Options/
      Alternatives/
      Variations 93
  7.1.3 Vacuum Belt Filter:
    Tray Type 94
    7.1.3.1 Slurry Feeding/
      Filtration 96
    7.1.3.2 Cake Washing 96
    7.1.3.3 Cake Pressing
      (Optional) 96
    7.1.3.4 Air Drying 97
    7.1.3.5 Cake Discharge 97
    7.1.3.6 Cloth Cleaning 97
    7.1.3.7 Installation 98
    7.1.3.8 Options/
      Alternatives/
      Variations 98
    7.1.3.9 Applicability 98
    7.1.3.10 Maintenance
      Notes 99
  7.1.4 Vacuum Belt Filter:
    Rubber Belt Type 99
    7.1.4.1 Slurry Feeding/
      Filtration 99
    7.1.4.2 Washing 99
    7.1.4.3 Air Drying 99
    7.1.4.4 Cake Discharge 100
    7.1.4.5 Cloth Cleaning 100
    7.1.4.6 Options/
      Alternatives/
      Variations 100
    7.1.4.7 Applicability 100
    7.1.4.8 Maintenance
      Notes 100
    7.1.4.9 Installation
      Notes 101
  7.1.5 Pan Filter 101
    7.1.5.1 Slurry Feeding/
      Filtration 101
    7.1.5.2 Washing 101
    7.1.5.3 Air Drying 102
    7.1.5.4 Cake Discharge 102
    7.1.5.5 Cloth Cleaning 102

Solid-Liquid Filtration. DOI: 10.1016/B978-0-08-097114-8.00007-1
Copyright © 2012 Elsevier Ltd. All rights reserved.

7.1.5.6 Options/Alternatives/Variations	102	7.3.3.7 Applicability	117
7.1.5.7 Applicability	103	7.3.3.8 Installation Notes	117
7.1.5.8 Maintenance Notes	103	7.3.3.9 Maintenance Notes	117
7.2 Pressure Filtration: Continuous	**103**	7.3.4 Candle Filters	117
7.3 Pressure Filtration: Discontinuous	**104**	7.3.4.1 Slurry Feeding/Filtration	118
7.3.1 Filter Press	104	7.3.4.2 Washing	118
7.3.1.1 Slurry Feeding	109	7.3.4.3 Air Drying	118
7.3.1.2 Cake Pressing	109	7.3.4.4 Cake Discharge	119
7.3.1.3 Cake Washing	109	7.3.4.5 Cloth Cleaning	119
7.3.1.4 Air Drying	110	7.3.4.6 Installation	120
7.3.1.5 Cake Discharge	111	7.3.4.7 Options/Alternatives/Variations	120
7.3.1.6 Cloth Cleaning	111	7.3.4.8 Applicability	120
7.3.1.7 Applicability	111	7.3.5 Spinning Disc Filters	120
7.3.2 Tower Press	111	7.3.6 Leaf Filters	121
7.3.2.1 Slurry Feeding	112	**7.4 Centrifugal Filtration**	**121**
7.3.2.2 Cake Pressing, Cake Washing and Air Drying	112	7.4.1 Batch Centrifuges	121
7.3.2.3 Cake Discharge	114	7.4.1.1 Slurry Feeding/Filtration	121
7.3.2.4 Cloth Cleaning	114	7.4.1.2 Washing	122
7.3.2.5 Options/Alternatives/Variations	114	7.4.1.3 Gas Drying	122
7.3.2.6 Applicability	114	7.4.1.4 Cake Discharge	122
7.3.3 Tube Press	115	7.4.1.5 Options/Alternatives/Variations	122
7.3.3.1 Slurry Feeding/Filtration	116	7.4.1.6 Applicability	123
7.3.3.2 Washing	116	7.4.1.7 Installation Notes	123
7.3.3.3 Air Drying	116	7.4.1.8 Maintenance Notes	123
7.3.3.4 Cake Discharge	116	7.4.2 Continuous Centrifuges	123
7.3.3.5 Cloth Cleaning	117	7.4.2.1 Cake Discharge	123
7.3.3.6 Options/Alternatives/Variations	117	7.4.2.2 Applicability	123
		7.5 Summary	**124**

A common mistake that people make when trying to design something completely foolproof is to underestimate the ingenuity of complete fools.

Douglas Adams, *Mostly Harmless*, 1992

The purpose of this chapter is to provide an introduction to the types of filters that are available. It is not intended to be the last word on equipment selection.

The information given here will not favor any particular design or individual manufacturer over any other. But the fact remains that some filter types are more suited for some applications and other types are wholly unsuited to some applications.

Within each filter type, individual manufacturers may offer their own unique features or subtle design changes that would make their particular machine most suitable for a particular application. It is beyond the scope of this book to go to this level of refinement: this is the job of the salespeople working for those manufacturers. Section 13.2 aims to give further insights into the process of equipment selection.

Industrial filtration equipment has been around for more than 100 years and evolution (through commercial selection) has given us a diverse selection of filter designs and a wide range of manufacturers from which to choose.[1]

This evolution has occurred with a strong following wind from constant innovation and technical development. This is especially true for cloth design, instrumentation/automation, manufacturing technology, design tools, standardization and materials technology. However, while reliability, size of filtration area and performance have increased, there have been few completely new slurry filtration devices during the past few decades. Many of the filters presented here closely resemble their ancestors from fifty years ago, albeit with higher precision in their manufacture, and more sophisticated control systems.

Suppliers of filtration machinery range from large, international players (usually with a number of machine types and technologies) through smaller, local equipment suppliers to, basically, workshops that can produce one-off machines to a given design. Each of these organizations employs designers with their own opinions, or prejudices, which add further to the variety of available designs.

Virtually all of the filters in this chapter have a cloth – the single most important component – that is supported by a grid of some sort to allow filtrate to flow. Some common cloth-supporting systems are:

- a grid, sitting in a filtrate tray (Figure 9.8)
- pips molded or machined onto a surface (Figure 7.23)
- a wedge-wire steel mesh
- a coarser cloth or felt.

If you zoom into the cloth/support part of any of these filters, it may be hard to tell them apart even if the operating principles and motive force for filtration used may vary. The other essential challenges of filter design remain:

- to distribute slurry onto a filter cloth, so that the motive force for filtration can be applied to the slurry

---

[1] Design ideas that do not offer a competitive advantage to filtration processes, in terms of cost, quality, safety, health and the environment, or productivity, tend not to survive.

- to contain the motive force for filtration (this is a particular difficulty for a continuous filter – slurry coming in must balance the cake and filtrate leaving, while the motive force for filtration is contained)
- to allow filtrate to flow, unimpeded, from the device
- to discharge the filter cake, either continuously or intermittently.[2]

There are many ways to categorize filtration devices. Figure 7.1 does this according to the motive force used for filtration:

- vacuum
- pressure
- centrifugal

and then mode of operation, i.e. whether the filter discharges filter cake:

- batchwise or
- continuously

because this will be a key consideration in process/plant design.

While this way of classification serves a purpose and covers the majority of cases, there are exceptions (e.g. some centrifuges can also be pressurized).[3]

7.1 VACUUM FILTRATION: CONTINUOUS

Many of the processes in the mining and chemical sectors, perhaps less frequently in the food and pharmaceutical sector, use continuous vacuum filters. These filters use a vacuum to provide the motive force for filtration. Slurry is fed continuously to the devices and filtrate and cake are produced continuously.

7.1.1 Rotary Vacuum Drum Filter

Figure 7.2 gives the overall construction of a rotary vacuum drum filter (or RVDF or simply vacuum drum filter). The drum is usually covered in a number of separate panels, containing a support grid, and these are covered in a cloth (and, often, a backing cloth). The cloth can either be fixed, or caulked,[4] into

---

[2] Usually to atmospheric pressure, but sometimes the next step in the process is above or below atmospheric pressure.

[3] Over the years, the terminology surrounding filtration technology has become a little fragmented, to the point where filter types with the same generic names from different manufacturers or from different market areas may be completely different. For example, a vertical pressure filter can be a pressure filter with a vertical stack of horizontal plates or a set of vertical plates arranged horizontally. Also, a vacuum disc filter bears no resemblance to a rotating disc-discharge filter. However, either may be known as the "disc filter" on a plant. Further confusion exists over the word "membrane", which can refer to an impermeable diaphragm used to squeeze a filter cake in a chamber filter or to a microfiltration, ultrafiltration or nanofiltration medium.

[4] The cloth is wedged into grooves using a rope that is hammered into place.

Chapter | 7 Filter Design

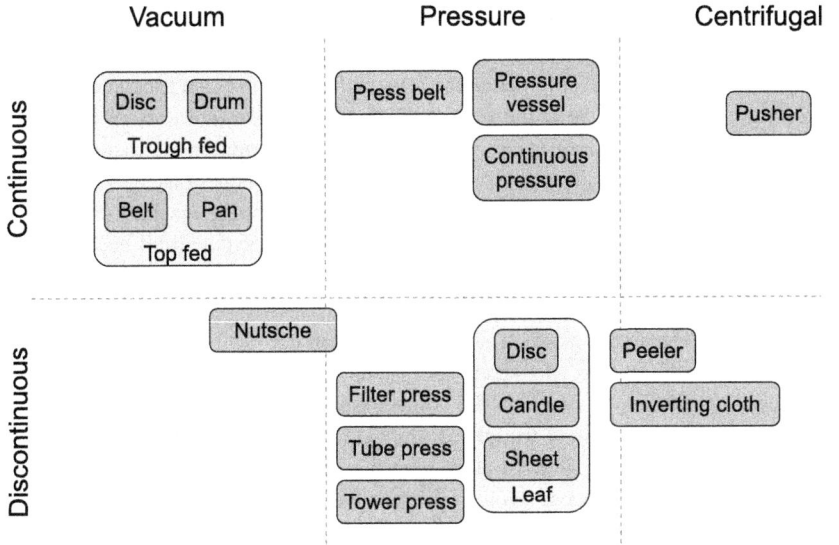

FIGURE 7.1 A categorization of filter types.

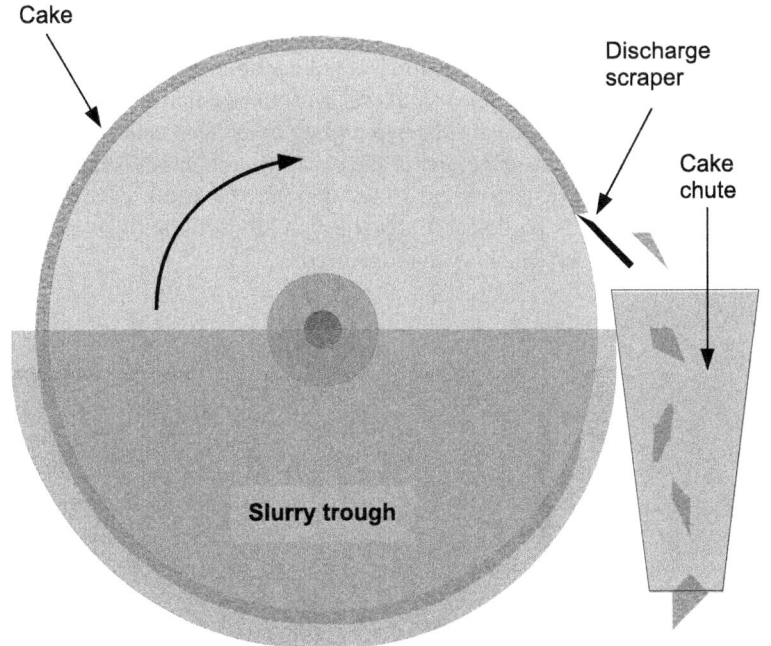

FIGURE 7.2 General arrangement of a drum filter.

panels that cover the entire surface of the drum or, less commonly, the whole drum can be covered in a one-piece cloth.

Each of these panels on the outer surface of the drum has a piped connection to a rotary control valve (Figure 7.3). The control valve (sometimes known as the control plate or rotary valve) is an essential component in all vacuum drum, vacuum disc and vacuum pan filters. Its purpose is to connect each of the points on the drum, in sequence, to a vacuum (for filtration, cake washing or air drying) then either to vent to atmospheric pressure or to give a small back-flow of pressurized air for cake discharge.

The drum is supported on bearings and rotated, either directly by an electrical motor or through a gearbox. During operation the drum rotates at a uniform speed (typically from 0.1 to 4 rpm) and each of the panels will, in turn, be connected to the different zones on the control valve.

The drum rotates in a trough of slurry. Typically, the slurry level in this trough comes to a little under halfway up the drum, as in Figure 7.2. However, since the filtration process cycle is limited in that it must take place entirely within this one rotation, there is relatively little flexibility in the cycle (it would not be possible to filter for ten seconds and then air dry for three minutes). For some relatively freely dewatering slurries, a much shallower slurry trough height can be used (together with appropriate changes to the location of the cake discharge in the control-valve plate).

7.1.1.1 Slurry Feeding/Filtration

In most vacuum drum filters, the slurry is fed continuously to the trough. The rate of feeding can be determined by a level sensor and control valve or an overflow weir (with the overflowing slurry returning to the feed tank).

As a panel becomes submerged in the slurry trough, it is connected to the vacuum system and, as the liquid flows through the cloth, cake formation begins. If the slurry is particularly coarse or rapidly settling, then an agitator may be needed to keep the slurry homogenized.

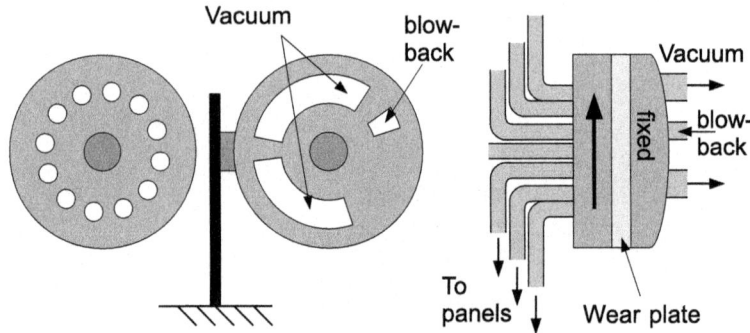

FIGURE 7.3 The control head: a key component in rotary vacuum filters.

Chapter | 7 Filter Design

In cases where the slurry filtration is extremely rapid, it is possible also to use a top-feed of slurry.

A rotary vacuum drum filter is shown in Figure 7.4.

7.1.1.2 Washing

Washing usually takes place on the back side of the drum (the opposite side to the cake discharge). The wash liquid can be poured onto the top of the drum, or possibly sprayed onto the cake on its way up (or even down).

Care should be taken that an appropriate amount of wash water is passed to the filter. Ideally, the water will fall down the surface filter cake in a uniform front and disappear into the cake just before it arrives at the free surface of the slurry in the trough. If the main purpose of the filter is to wash the filter cake, then any dry cake surface on the back of the filter is a lost opportunity. Equally, if too much wash water is applied then it can cascade down the cake and into the feed trough, where it merely dilutes the slurry.

7.1.1.3 Cake Pressing (Optional)

Although it is relatively uncommon, it is possible to press the cake during its rotation to prevent cake cracking or to encourage more moisture removal. This can be in the form of either a mechanical pressing device like the type shown in Figure 7.5, or a seal belt, illustrated in Figure 7.6(d), which uses atmospheric pressure.

FIGURE 7.4 Rotary vacuum drum filter, assembled ready for delivery (FLSmidth).

FIGURE 7.5 Rotary vacuum drum filter with press belt (FLSmidth).

7.1.1.4 Air Drying

The remainder of the time on the drum is used to draw air through the cake to remove moisture. As discussed above, there is a limit to the range of relative times that can be selected for air drying and filtration. It is possible to use a hood to direct hot air over the cake to help cake drying, although there would be a limit to the effectiveness of this arrangement at high speed.

7.1.1.5 Cake Discharge

A number of alternative cake discharge mechanisms are available (see, for example, Figure 7.6). The basic scraper (a) works well when the cakes are relatively thick and non-sticky. Where the cake is sticky, or thin, then alternatives can be installed. In the roller discharge system (b), the cake is peeled away from the cloth by a roller that rotates in the opposite direction to the main drum, then the cake is scraped from this roller, either by a scraper blade or by a set of combs. In the cloth discharge alternative (c), the drum is covered with a single cloth, which is taken away from the drum and the cake discharged as it passes over a smaller diameter roller. Another advantage of this system is that the cloth can be washed (on both sides, if necessary) before it returns to the drum. A similar arrangement

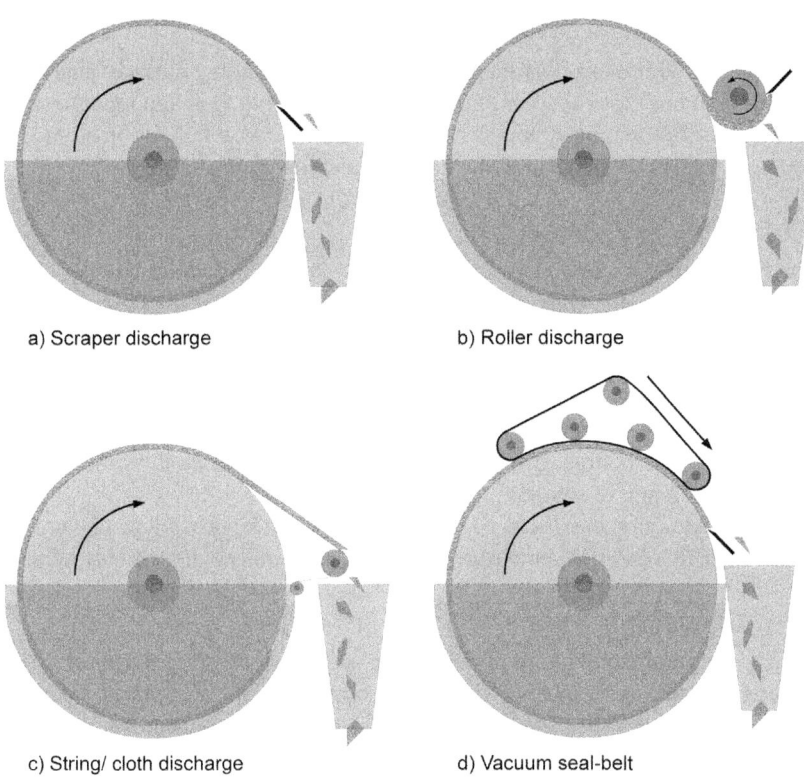

a) Scraper discharge   b) Roller discharge

c) String/ cloth discharge   d) Vacuum seal-belt

FIGURE 7.6 A variety of drum filter configurations.

to (c) has a set of strings that cover the drum and bring the cake away from the drum (in this case, a conventional cloth arrangement can be used).

7.1.1.6 Cloth Cleaning

If the cloth is fixed to the drum panels then the choices for washing while the filter is online are limited. High-pressure spray bars can be used intermittently, and during offline washing it is possible to back-flush wash liquid through the filtrate system and back through the filter cloths.

If the filter cloth needs to be cleaned continuously, then the arrangement shown in Figure 7.6(c) can be used.

7.1.1.7 Installation

As with all filters with the cake exposed, it is useful if the process can be seen. If the wash side of the filter is hidden away from a decent viewing platform or bathed in clouds of steam, it will be difficult to spot blocked nozzles or fouled wash distribution weirs.

7.1.1.8 Maintenance Notes

In common with other continuous filters, there are sliding surfaces that need attention, in this case the wear plate in the control valve. It is important to make sure that the lubrication system is in good order, otherwise failure can be sudden and potentially expensive in terms of repair and lost production.

7.1.1.9 Options/Alternatives

For slurries that are very easy to filter, a thick cake may form in a few seconds and, in this case, the slurry level can be much lower. It is possible to have a slurry filtration zone that covers only around 60° of the rotation, with the bulk of the remaining cycle used for air drying.

Where the slurry is particularly fast settling (or the throughput of liquid not rapid enough to suspend the solids), an agitator may be needed to prevent solids falling to the bottom of the trough (and damaging the cloth as the drum is scraped over a pile of solids).

Fume extraction hoods can be installed, and it can be possible to enclose the entire filter in a gas-tight enclosure if the process requires an inert atmosphere.

7.1.1.10 Applicability

Rotary vacuum drum filters operate in a wide variety of applications, from general dewatering of mineral slurries to complex washing applications in the chemical industry. They have existed for around 100 years and continue to be developed.

7.1.1.11 Precoat Filters

Vacuum drum filters can operate as precoat filters, in which a cake of precoat is formed on the drum and then slurry (usually with a low solids content) filtered to produce a very clear filtrate. In operation, a thick cake of precoat material (e.g. diatomaceous earth or perlite; see Section 3.3.5) is first laid down on the drum before the liquid to be filtered is applied to the drum (usually via the trough, but occasionally top-fed).

The drum filters used in this way resemble conventional vacuum drum filters, although there are two crucial differences. First, a very high precision scraper is used to shave away very thin layers of cake, either continuously or intermittently. This blade advances slowly towards the cloth and, since the thickness of the precoat bed can be more than 100 mm and the speed of the blade can be of the order of millimeters per hour (or micrometers per drum rotation), a unit can, potentially, be operating for many hours, or perhaps days, on a single charge of precoat.

Second, the control valve does not have a vent or blow-back zone, because the cake must be kept under vacuum for the whole of the drum rotation. (It is possible for the control valve to have a separate zone that is kept under vacuum under normal operation, but switched to vent/blow-back to discharge spent precoat.)

This type of filter is often used to remove small concentrations of slimy, fibrous or gelatinous solids (that would otherwise clog a filter cloth) or where very high filtrate clarity is needed. In many circumstances, because of the shaving away of a thin skin of used precoat, this offers a low consumption of filter aid, since each layer is made to work fully. It is also possible to use a reactive precoat medium, for example activated carbon for the decolorization of glucose syrup.

Conventional vacuum drum filters can be retrofitted with the necessary auxiliary equipment to operate as precoat filters, although the precision and stability of the drum will need to be checked.

The speed of the scraper knife will determine the throughput of the filter and the unit can be controlled on this basis (if the flow rate drops, the speed of the blade travel can be increased). It is very difficult to mimic this type of filtration in the laboratory (although a quantitative assessment can be made) and pilot-scale testing under very carefully controlled conditions will nearly always be necessary.

7.1.2 Rotary Vacuum Disc Filter

The vacuum disc filter operates in a similar way to the vacuum drum filter: filtration elements (sectors), covered in a filter cloth, pass into a slurry trough, vacuum is applied to the sector, a cake forms on its surface, it emerges from the slurry and the cake is dried by air passing through it. Finally, a back-pulse of air to the inside of the sector dislodges the cakes (which fall into a chute that passes through the slurry trough), before the sector passes back into the slurry trough and the process is repeated.

Rotary vacuum disc filters have some key differences from rotary vacuum drum filters:

- The slurry level in the trough must be higher than the top of the sectors as they pass through the trough (otherwise air would simply pass through the cloth during cake formation).
- The filtration area (and therefore throughput) can be higher than a drum filter for the same floor space.
- Vacuum disc filters are less commonly used for cake washing (because of the vertical cake surface and the difficulty in getting the wash liquid onto the cake). Where washing is used, a system of spray nozzles is normally used to flood the surface of the cake with a mist.
- There are fewer cake discharge options for sticky cakes.

Figure 7.7 shows a side view of a typical disc filter.

The filter cloth is usually supplied as a cone and is fitted over the sector and sealed top and bottom. Each sector is connected to the control plate by individual pipes (or hoses) or through fabricated channels in the central barrel to form a disc. A number of discs can be fitted to the same central barrel.

The barrel is supported on bearings and may be driven either directly by a motor or indirectly through a gearbox.

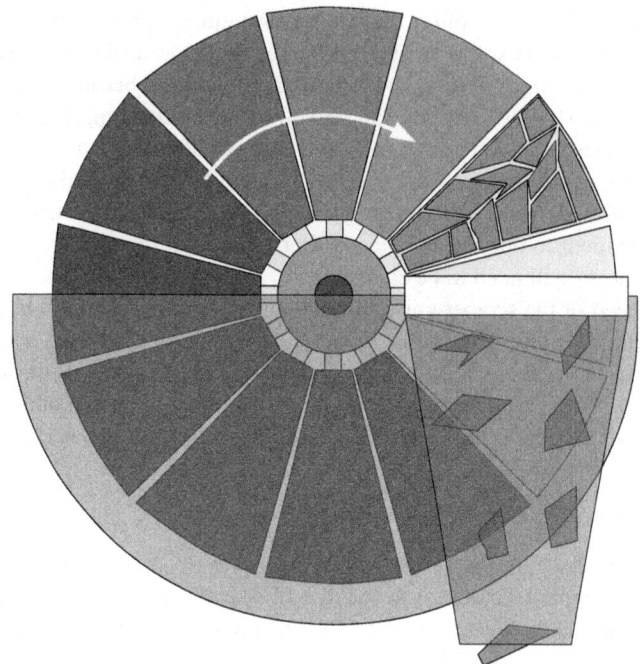

FIGURE 7.7 General arrangement of a vacuum disc filter.

The discs rotate in a slurry trough. This trough may be partly compartmentalized to reduce the volume held in it at any one time, and therefore reduce the residence time of slurry in the trough.

The slurry is normally fed directly into the slurry trough. The slurry feed can either be controlled by a feed valve/level sensor or be set at a standard rate (if the slurry composition, vacuum and cloth condition remain constant then so will the throughput) with a weir overflow sending surplus slurry back to the feed tank.

Initial cake formation and filtration occur during the sector's time in the trough.

7.1.2.1 Washing

Cake washing can be performed on a vacuum disc filter, once the sector has emerged from the trough. However, since the cakes are vertical, the wash liquid must be sprayed onto the cake surface. It is less common to use disc filters for washing than, say, drum, pan or belt filters.

7.1.2.2 Air Drying

Air drying takes place from the time that the sector emerges from the slurry and continues until cake discharge; this is usually about one-third of the rotation time (the equivalent of about 120°).

7.1.2.3 Cake Discharge

After air drying, the sector experiences a sharp back-pulse of air. This expands the cloth that covers the sector and the cake falls into the cake chute. In some cases, simply removing vacuum from the sector (using a blocking bridge in the control valve) will be enough.

Most larger disc filters do not use a scraper that touches the cloth, since the disc sectors are cantilevered out from a central barrel and the tolerance needed, on the scale of millimeters, is not possible. Any loose-fitting cloths could snag on a scraper.

7.1.2.4 Cloth Cleaning

Given the way that the filter operates and the short time between cake discharge and re-entering the slurry trough, it is difficult to clean the cloths while the filter is online. If the bags become clogged or blinded over time, it may be necessary for a periodic clean, perhaps in acid or caustic, maybe daily or weekly.

7.1.2.5 Applicability

Vacuum disc filters are used for relatively coarse slurries. Prominent applications are in hydrate seed filtration on Bayer process alumina plants or coarse metal ore concentrates, for example iron or copper.

7.1.2.6 Installation Notes

With larger disc filters, the individual sectors can be quite heavy and so a good platform, incorporating an easy-to-use crane or hoist, is a useful investment.

7.1.2.7 Maintenance Notes

As noted for drum filters, it is important to make sure that the lubrication system on the control valve is in good order. It is also particularly important to check the condition of the cloths, since one cloth failure can, in time, cause failure of other cloths and solid material finds its way into other sectors.

7.1.2.8 Options/Alternatives/Variations

Expansion of a disc filter is technically possible, by adding more discs. In most cases, however, vacuum disc filters are delivered with their full complement of discs already installed.

Ceramec™ capillary action disc filters are a particularly interesting variation on the disc filter principle. In nearly all respects, these filters look identical to conventional disc filters. However, the sectors are made of a microporous ceramic membrane and the capillaries in this membrane allow water but, crucially, not air to pass through. This means that the large vacuum pumps needed by conventional vacuum filters (which would draw a large amount of air through the cake) are not needed, giving a significant reduction in power consumption.

Furthermore, the microporous membrane provides extremely clear filtrate. Ceramec filters offer very significant benefits to some applications, but the range of applications is quite limited.

Ceramec disc filters use a finely adjusted scraper for cake discharge, since discs can be made to rotate within very tight tolerances (and there is no cloth to catch or snag on the scraper).

7.1.3 Vacuum Belt Filter: Tray Type

Tray-type vacuum belt filters have been developed since the 1960s and offer extremely flexible and configurable filtration process choices. In these filters, the cloth is acting not simply as the filter medium, but also as the cake conveyor.

There are two main types of construction and operating principle for tray-type belt filters, both of which involve a filter cloth moving over a set of filtrate trays containing a support grid (Figure 7.8). These two basic types are:

- reciprocating tray
- stop–go.

A reciprocating tray filter comprises a continuously moving cloth and a set of trays on wheels, each one connected to a common manifold with flexible hoses. Once a vacuum is applied to these trays, the cloth is sucked down and the trays will follow the cloth, drawing filtrate and wash filtrate as it goes. Before the trays collide with the cloth roller at the discharge end of the filter, there is a brief pause while the vacuum is cut and the manifold vented to atmosphere. At this moment, the trays are drawn back sharply to the beginning and the process is repeated. In practice, the aim is for the trays to be under vacuum for most of the time. The operation of a reciprocating tray-type filter is shown in Figure 7.9.

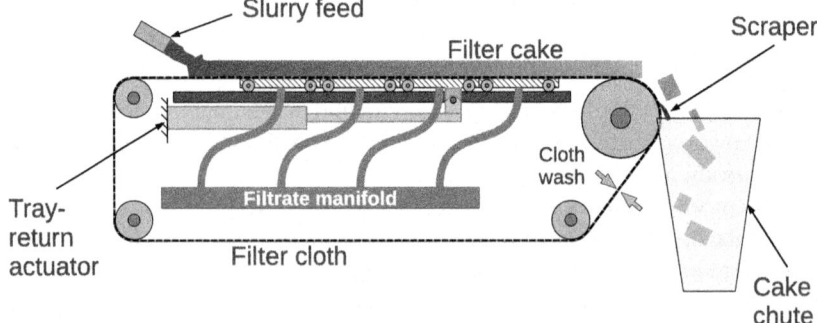

FIGURE 7.8 General arrangement of a tray-type belt filter, cloth tensioning and cloth tracking omitted. In reciprocating tray filters, the trays move, and are returned by the actuator shown. For stop–go filters, the trays are fixed.

Chapter | 7 Filter Design

A stop–go type of filter is extremely similar in principle but, as the name suggests, the cloth advances a certain distance before the vacuum is applied and the filtration, washing and air-drying processes take place.

There may be particular reasons why one of these types could be particularly suitable (or unsuitable) for a particular application, but this will require thorough investigation during equipment selection.

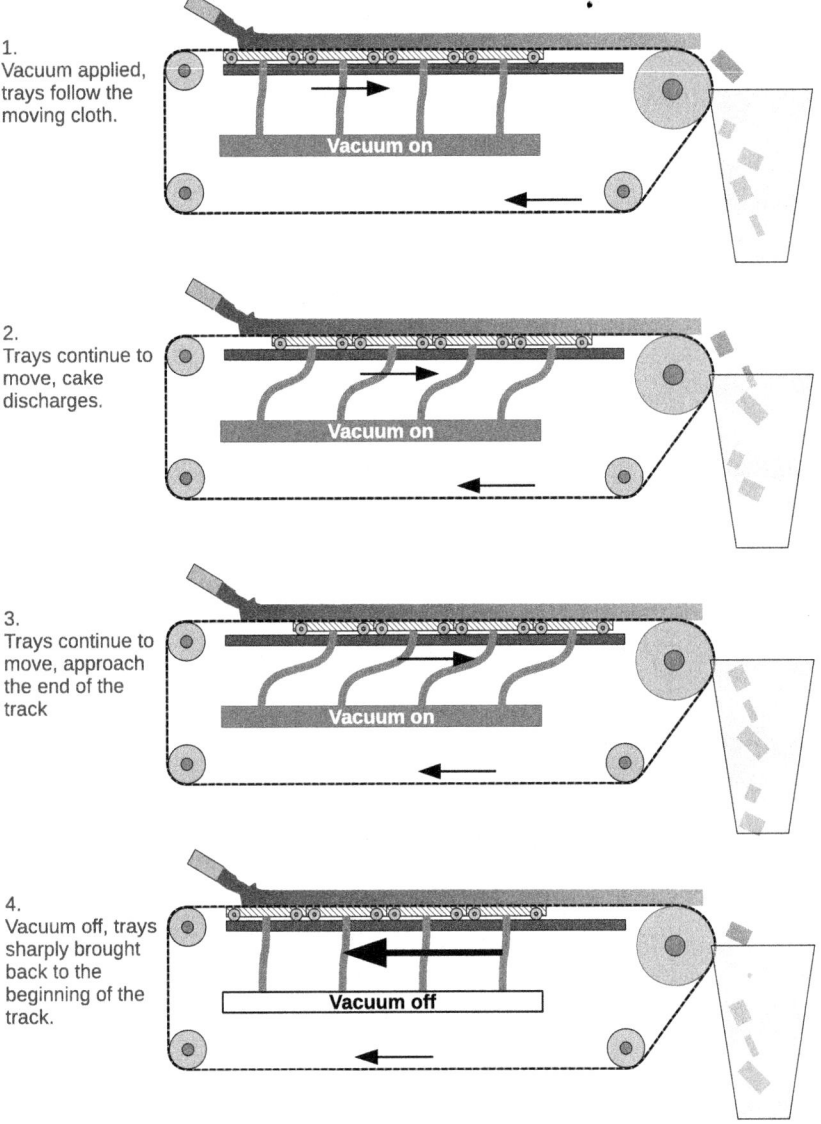

FIGURE 7.9 Reciprocating tray belt filter.

In both cases, the cloth is a single piece and the two ends are joined, usually with a zipper seam or gluing/welding.

7.1.3.1 Slurry Feeding/Filtration

Slurry is fed directly onto the belt, usually through a manifold with a number of ports, or a feed weir/plate to distribute the slurry evenly over its width. For reciprocating tray filters, this feed is continuous and through a fixed feeding arrangement. For stop–go type filters this slurry feed may itself reciprocate to cover a step-length of cloth before the vacuum is applied.

7.1.3.2 Cake Washing

A huge number of processing options are available on belt filters. It is possible to design the duration of the various stages of a filtration cycle by selecting the appropriate position on the belt.

It is also very feasible to have a large number of washing stages, including counter-current and reflux washing options (see Sections 3.2.5.1 and 3.2.5.2). Figure 7.10 shows a belt filter with multiple washing stages installed. In certain processes, the cloth wash water can be reused for cake washing.

7.1.3.3 Cake Pressing (Optional)

Although relatively uncommon, it is also possible to include a cake-pressing stage into a belt filter. This can be either a simple seal belt (as discussed in Section 7.1.1.3) to consolidate a cake (under atmospheric pressure) and prevent

FIGURE 7.10 Belt filter with multiple washing stages (BHS Sonthofen).

air flow through it before a washing stage or, as before, a more elaborate mechanical or intermittent bladder pressing device.

7.1.3.4 Air Drying

One benefit of belt filtration is that the belt can be made as long as is needed for air drying (within certain practical limits). Air drying can be particularly effective given the uniform cake properties that belt filters can provide (with fewer opportunities for bypassing of drying air).

It is possible to apply hot-air drying stages to vacuum belt filters.

7.1.3.5 Cake Discharge

The cloth usually passes over a roller at the end of the filter and, for some thicker, drier cake, this action alone is enough to send the cake tumbling down the cake chute. Thinner, stickier cake may need either a scraper blade or, sometimes a series of wire scrapers to persuade the cake away from the cloth (Figure 7.11).

7.1.3.6 Cloth Cleaning

The cloth on a belt filter is cake-free for about half of the time, and this time can be used to clean the cloth thoroughly. Figure 7.8 shows one possible configuration. However, it is also possible to pass the cloth through a series of chemical or ultrasound baths to remove any persistent solids from the cloth.

FIGURE 7.11 Belt filter cake discharge (Outotec Filters).

Depending upon the needs and limitations of the process, it is possible to return the cloth wash liquid (usually water) to the process (or to use it for cake washing) so that virtually no solids are lost.

7.1.3.7 Installation

One of the major benefits of a belt filter is that the process is there to be seen, moving at a reasonable rate, so it is possible to monitor exactly how it is going and make any necessary adjustments to the filter operation. It is important to make sure that the installation allows this to happen, with the provision of good lighting and ventilation. One advantage of belt filters is that they can be installed at floor level, so often there is no need for a special viewing platform.

7.1.3.8 Options/Alternatives/Variations

Using hot gas, infrared or even microwave drying stages along the length of a belt filter, it is possible to discharge powder-dry material from the end of a belt filter. Figure 7.12 shows an example of a belt filter within a fume hood.

There are also alternative constructions to house the filters, either in a gas-tight enclosure to prevent volatile emissions, ensure hygiene or bathe the process in (for example) nitrogen gas, or in pressure-tight enclosures so that overpressure can be used as a force for filtration. (see Section 7.2, on continuous pressure filters.)

7.1.3.9 Applicability

Tray-type belt filters are widely used. The wide range of configurations and options means that they tend to find their place in applications where their washing and drying capabilities can be exploited.

FIGURE 7.12 Belt filter, showing fume hood (Outotec Filters).

7.1.3.10 Maintenance Notes

Tray-type belt filters are notable for the long lifetime of their spare parts (especially cloths) and, because of their simplicity and rather slow pace, their good reliability and availability.

7.1.4 Vacuum Belt Filter: Rubber Belt Type

The rubber belt filter is a close cousin of the tray-type belt filter, and consists of a continuously moving grooved rubber belt acting as a cloth support, filtrate collector and cake conveyor (Figure 7.13).

The belt itself can be a relatively complicated structure with a series of grooves either machined or molded onto its surface that function as filtrate collectors. These grooves drain through holes, often, but not always, arranged along the center-line of the belt into a vacuum box that is pressed against the underside of the belt.

The belt can be made from many different types of rubber, for chemical compatibility, and includes reinforcement.

The surface of the vacuum box that the belt slides over must be lubricated (usually with water). The consumption of this water can be a significant cost factor.

7.1.4.1 Slurry Feeding/Filtration

Since the belt is moving continuously, the slurry can be fed to the filter at a constant rate.

7.1.4.2 Washing

The vacuum box can be compartmentalized so that the wash filtrates from different compartments can be used for counter-current or reflux washing as for tray-type filters.

7.1.4.3 Air Drying

Once again, the process can be designed to have an extremely long (or short) air-drying time if necessary.

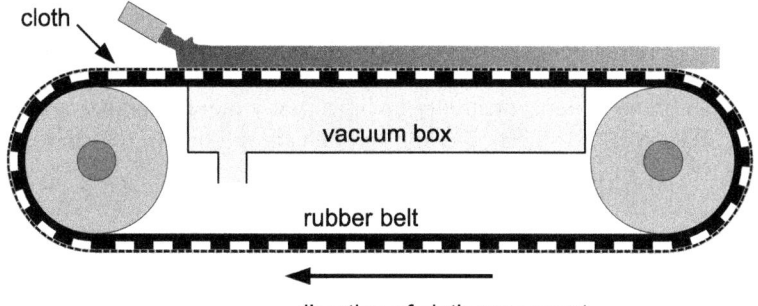

FIGURE 7.13 General arrangement of a rubber belt filter.

7.1.4.4 Cake Discharge

As with other belt filters, cake discharge can be with a simple scraper or sometimes, particularly for a sticky or very thin cake, with wires or other discharge systems. The discharge is continuous.

7.1.4.5 Cloth Cleaning

The cloth can be cleaned as it passes a fixed washing station (usually high-pressure spray bars). Depending upon the application, this cleaning can be continuous, occasional or even controlled automatically (e.g. by measuring air flow through the belt and washing once this flow drops). The dirty cloth wash water can also be passed onto the cake to minimize the loss of solid material from the process.

If the cloth stays against the belt along its whole length (as it does in Figure 7.13) then the spray bars can only wash one side of the cloth directly (and in the same direction as the filtration process). Other arrangements can bring the cloth away from the belt, through a system of rollers, so that it can be washed on both sides or even passed through a chemical bath.

7.1.4.6 Options/Alternatives/Variations

As with other belt filters, the units can be installed with fume extraction hoods, through gas-tight configurations and even pressure-tight enclosures.

7.1.4.7 Applicability

Rubber-belt vacuum filters are commonly used for very large-scale mineral and mining dewatering applications. In particular, this type of top-fed filter is useful in applications with rapidly settling slurries. For some rapidly settling materials, it may be possible to classify the slurry into relatively coarse and fine fractions and use two stages for feeding, with the coarser fraction being laid onto the cloth, like a precoat, before the finer fraction is added further along the belt. This scheme would protect the cloth and produce a clearer filtrate.

Since the cloth itself is not doing any cake conveying by itself, the same cloth may have a slightly improved lifetime compared with a tray-type belt filter. However, if the cloth is firmly fixed in one location on the belt, alternate bands of the cloth will have filtrate passing through at comparatively higher and lower rates, potentially giving slightly more wear and tear to the bands that have the filtrate passing through more quickly into the groove space.

7.1.4.8 Maintenance Notes

Provided that the belt/vacuum box interface is lubricated and other key parts of the machine are kept clean (bearings/motors/seals), machine reliability can be very good.

Overwhelmingly, the most expensive component is the belt itself. Although its lifetime can extend for many years, replacement of the belt is also a significant operation. If you have a number of filters in operation, it makes sense to keep a replacement belt in store. Otherwise, the lead-time for a replacement belt can be quite long, so the condition of the belt should be closely monitored.

7.1.4.9 Installation Notes

On a rubber belt filter the process proceeds in front of your eyes – if you can see the belt. A viewing platform at the appropriate level will let the operators who are responsible for the process see what is happening, so if there is a significant process deviation it will be spotted and steps can be taken to fix the issue.

It is also important that the belt-changing process is understood and is possible within the arrangement.

7.1.5 Pan Filter

Also known as the table or horizontal filter, the pan filter resembles a rotary vacuum disc filter laid on its side. The pan is compartmentalized into trays, each containing a support grid and cloth arrangement. Each of these trays connects to a control valve in the center of the pan.

From an operational point of view a pan filter functions like a belt filter, although the operations occur in sequence during one rotation rather than along the length of a belt.

7.1.5.1 Slurry Feeding/Filtration

Slurry feeding occurs at the beginning of a rotation (Figure 7.14) and some care needs to be taken to ensure that the distribution of slurry is uniform, with less slurry fed onto the cloth near to the inner edge of the pan.[5] There will be some self-correction but, particularly if the slurry is fast filtering, the distribution can affect the uniformity of the cake.

This is important because variation in cake thickness (and other properties) can affect the rate at which wash liquid and drying air pass through the cake, preferring to flood through thinner regions.

7.1.5.2 Washing

Pan filters are commonly used on washing application, for similar reasons to belt filters. Some care must be taken to make sure that the distribution of wash water is appropriate over the width of the disc (as for the slurry distribution). Wash weirs or spray nozzles are normally used to apply the wash liquid directly onto the cake.

---

[5] This is because the further away from the center of the pan, the greater the area per unit width.

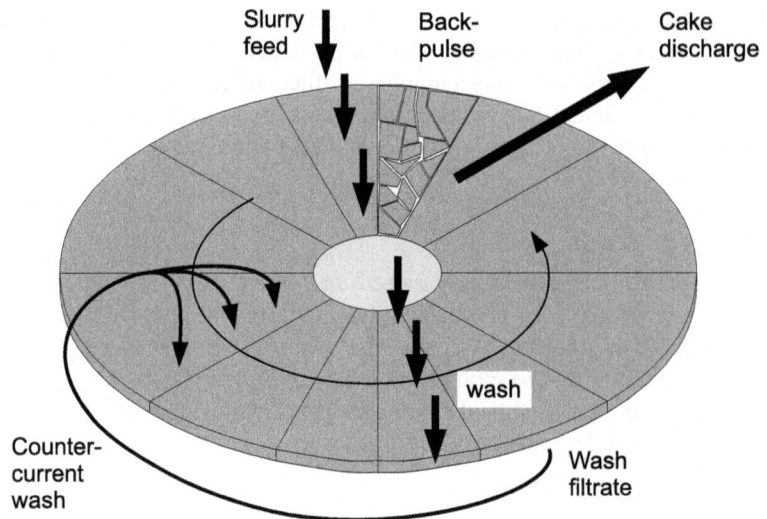

FIGURE 7.14 General arrangement of a pan filter.

If the control valve is compartmentalized then wash filtrate from one point on the rotation can be returned to an earlier point for counter-current washing. However, there is a practical limit to the number of stages that can be used.

7.1.5.3 Air Drying

As with other rotary vacuum filters, air drying occurs once filter cake formation and any washing stages have been completed.

7.1.5.4 Cake Discharge

Usually, a discharge screw conveys the cake sideways from the pan into a cake chute. In this arrangement, the discharge screw does not actually touch the filter cloth, usually leaving a 20 mm heel of filter cake.

An alternative is an arrangement in which the individual sectors on the filter turn to dislodge the filter cake. This is known as a tilting pan filter.

7.1.5.5 Cloth Cleaning

After cake discharge, there is often a back-pulse of air to disrupt the heel (the cake that passes under the discharge screw). This may be accompanied by sprays of water to reslurry this cake.

7.1.5.6 Options/Alternatives/Variations

Given that these filters often perform washing duties with hot wash water, they are often supplied with extraction hoods to draw away steam.

7.1.5.7 Applicability

Because the cloth is not acting as a conveyor, as it is on, say, a tray-type belt filter, the cloths can be selected solely for their filtration performance. Also for this reason, pan filters can handle high temperatures, for example those using very hot condensate for washing.

7.1.5.8 Maintenance Notes

As with other rotary vacuum filters, it is important to make sure that the control head and wear plate are well lubricated.

Extra care should be taken to make sure that nothing other than slurry and wash liquid go onto the cake. A small piece of trash, say a bolt-head, or even a piece of hardened filter cake, can catch under the discharge screw and rip a cloth.

In addition, it is tempting for people to walk over the filter cloth during other maintenance. A small piece of gravel stuck to the sole of a workboot can potentially weaken or even puncture the cloth. The result will be a pinhole that can rapidly expand and spoil the filtrate clarity and washing/air-drying performance of the process. So, feet on cloths should be avoided at all costs.

7.2 PRESSURE FILTRATION: CONTINUOUS

Most continuous filters installed to date use vacuum as the driving force for filtration. However, other choices offer continuous (or effectively continuous) slurry feeding and cake discharge operating under pressure. These offer the potential benefits of higher throughputs, better washing performance and lower final cake moistures together with that, often crucial, continuous operation.

These filters are relatively uncommon and will be discussed briefly here.

There are three main categories of continuous pressure filter:

- Filters that resemble continuous vacuum filters: These are housed in a pressure vessel, with slurry fed continuously into this housing and cake discharged through an airlock arrangement.
- Belt presses: In these, slurry is fed between two cloths that pass through a series of rollers that apply increasingly more force to the slurry. The cloths may also move at slightly different speeds to shear the cake. Cake washing and air drying are not normally applied to this type of dewatering device.
- The purpose-built rotary pressure filter manufactured by BHS Sonthofen.

Figure 7.15 shows a belt filter in a pressurized housing, but drum and disc filters may also be housed in a similar arrangement. In this example, the air used to pressurize the housing (or nitrogen if the atmosphere needs to be made inert) is separated in a filtrate receiver and recycled into the process (via a booster pump). Crucially, the cake discharge must take place through an airlock of some sort. In this case, a rotary valve is used, although two knife-gate valves in a

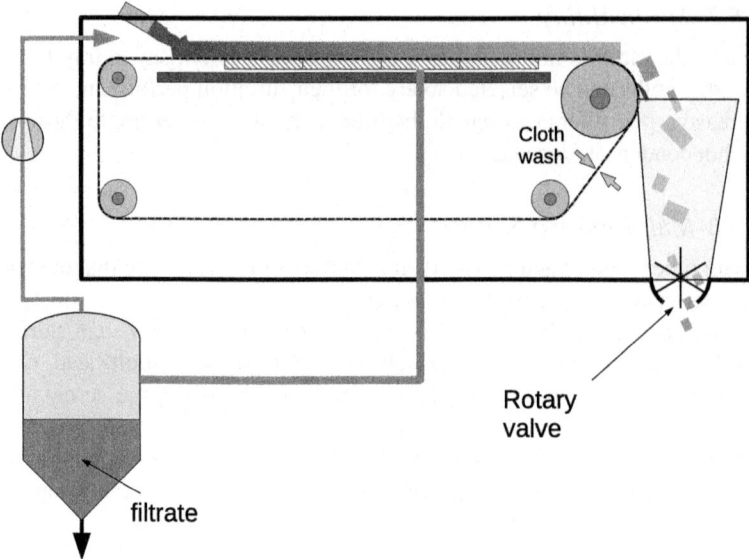

FIGURE 7.15 Example of a belt filter in a pressurized enclosure.

chute would also serve the purpose. If pressurized nitrogen is used, then a certain amount will be lost with the cake discharge.

In the rotary pressure filter (Figure 7.16), a central drum with a number of raised sealing surfaces rotates inside a drum so that the sealing surfaces form a number of moving chambers that pass through a number of zones. Slurry, wash liquid and drying gas are supplied under pressure to each of these zones. Finally, the cake is discharged through a chute.

7.3 PRESSURE FILTRATION: DISCONTINUOUS

All of these filters share the fact that they:

- close to form a pressure vessel of some sort
- open to allow cake discharge.

There are two broad categories, those filters that form individual filtration chambers and pressure vessels that contain a number of filtration elements. The first category includes filter presses, automatic tower presses and tube presses, while the second includes leaf, spinning-disc and candle filters.

7.3.1 Filter Press

The filter press is a general-purpose technology that can be used in a huge number of applications. This wide variation in application is reflected in the variety of designs, options and sizes available. The filter shown in

Chapter | 7 Filter Design

FIGURE 7.16 Continuous pressure filter: all stages (feeding/filtration, washing and air drying) take place under pressure, giving a continuous cake discharge (BHS Sonthofen).

Figure 2.1 shows a 100-year-old filter press. Filter presses comprise three main elements:

- a frame, or structure, that resists the pressure and maintains the structural integrity of the machine when it is under pressure
- a pack of filter plates, each covered in cloth (there can be more than 200 plates in a single filter unit)
- mechanisms for the opening and closing of the plate pack, cake discharge, cleaning, etc.

Figures 7.17 and 7.18 show two variations on the basic frame structure: side-bar and overhead beam. Each offers certain benefits, depending upon the application, but the overhead beam offers easier access to the plates for cake discharge and cloth changing.

In each case, the plate-pack opening and closing is controlled by high-pressure hydraulics, shown clearly in Figure 7.19. The structural end plates and the beams that stretch the length of the filters must be designed to withstand enormous forces once the plate pack comes under pressure.

Figure 7.20 shows a simple arrangement in which alternate frames and plates (covered in cloth) form a series of discrete filtration chambers, when pushed together. However, there is a huge variety of alternative plate types, for example:

- Chamber plates: Recessed trays in each plate come together to form chambers (in effect, the frame is incorporated into the plate).

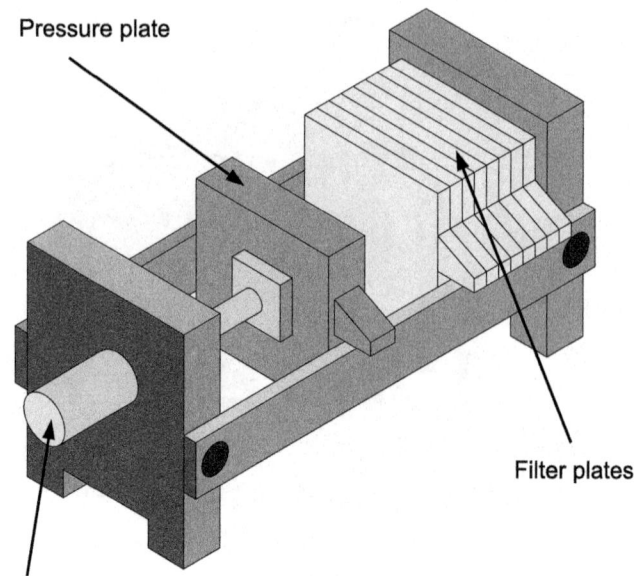

FIGURE 7.17 General arrangement of a side-bar filter press.

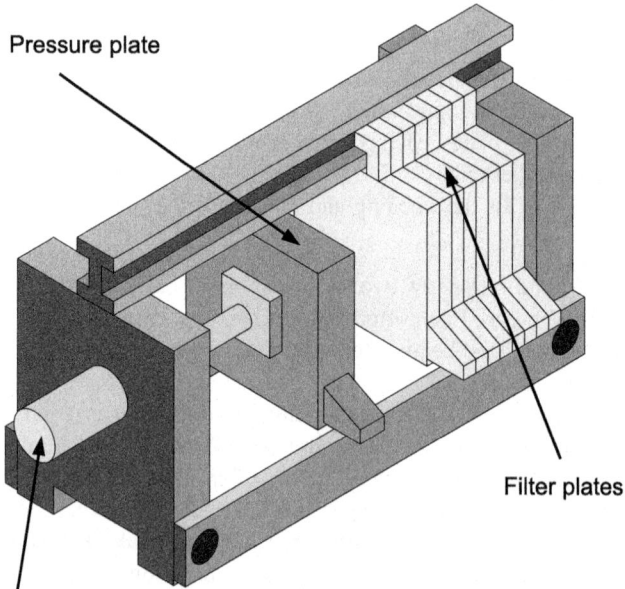

FIGURE 7.18 General arrangement of an overhead beam filter press.

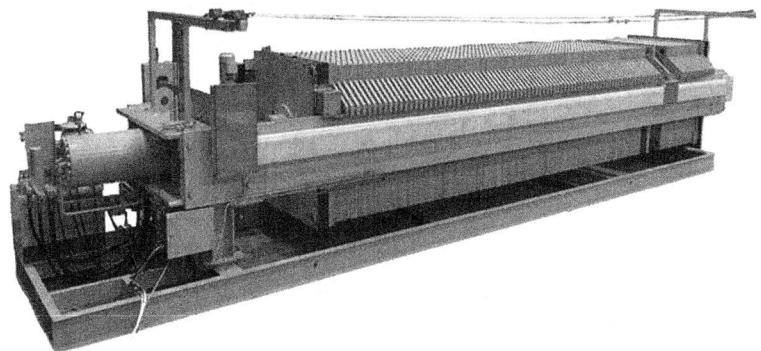

FIGURE 7.19 A side-bar filter press (Outotec Filters).

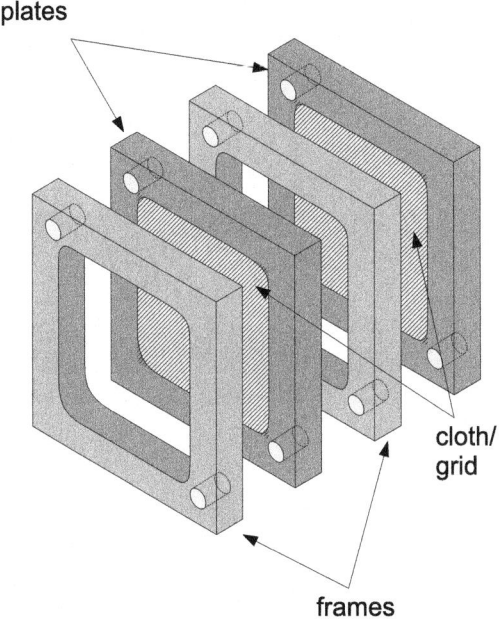

FIGURE 7.20 Example of a plate and frame pack.

- Diaphragm plates: The plate incorporates an impermeable diaphragm for cake pressing. This diaphragm can be a clip-in rubber or polymer membrane or can even be incorporated into the plate itself. The diaphragm may press directly onto the filter cake, or it may incorporate drainage pips and be covered in cloth, so that the cake can be pressed, with filtrate draining through the cloth, from both sides (Figure 7.21).

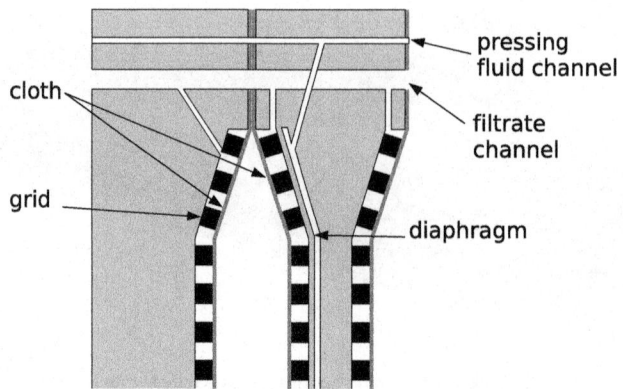

FIGURE 7.21 Detail of a diaphragm plate (r) with a recessed-chamber plate (l).

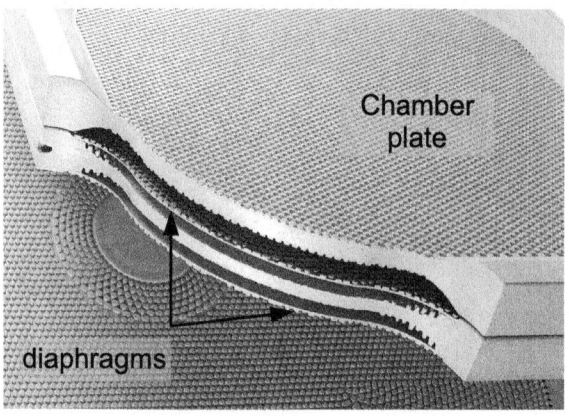

FIGURE 7.22 Diaphragm plate (with a chamber plate), showing the cloth support pips (Klinkau Filtration Systems).

In addition, the plates can be mixed (Figure 7.22). If alternative recessed chamber/diaphragm plates are used then cake pressing is applied from only one side of the cake.

In the early days of pressure filtration (see Figure 2.1), the plates could be made from hardwood, with drainage grooves machined into the wood. Over the past fifty to sixty years, molded polymer construction, mostly using polypropylene, has almost entirely taken over for most duties (although cast iron and aluminum plates are still used). With cast or molded plates, the cloth support grid can be cast into the plate itself (usually as an array of pips) (Figure 7.23).

In operation, the pack of plates is squeezed together to form a number of filtration chambers, each containing a filtration surface. The frame must keep the plate pack closed during filtration operations (the forces can be very significant) and then open to allow the filter cake to discharge.

Chapter | 7 Filter Design

FIGURE 7.23 A chamber plate, showing the cloth support pips, stay-bosses and filtrate drainage (Klinkau Filtration Systems).

7.3.1.1 Slurry Feeding

Slurry may be fed to each individual chamber through hoses, but normally the plates incorporate circular holes which, when all pushed together, form a pipe running the length of the plate pack. Slurry feeds from this pipe to each of the chambers (see Figure 7.24).[6]

Filtrate passes through the cloth and may pass through filtrate channels formed in a similar way to the slurry channels (holes lining up to form pipes).

7.3.1.2 Cake Pressing

Pressurized fluid, usually water or air, can be passed to diaphragms in each of the chambers so that they press the cake.[7] This pressing, normally at up to 16 bar, can remove a great deal of moisture and, if the slurry/cake is squeezed by cloth on both sides, there can be a considerable shortening of the time needed for this stage.[8]

7.3.1.3 Cake Washing

Cake washing can be arranged in all manner of different ways. For example, wash liquid can be flooded into the chamber and pressed through with a diaphragm again, or it can even be passed through the filtrate channels to flow from

---

[6] Where the plates come together to form channels (e.g. so that slurry can be passed through the plate pack to each individual filtration chamber), it may be necessary to have rubber inserts stitched or glued into the cloths in line with the slots in the plate pack.
[7] Air can flow more quickly into the chambers and may be used if a very short cycle time is needed, otherwise water usually offers a lower cost alternative.
[8] Presses that operate at 30 bar are becoming increasingly common, and some highly specialized machines can go even higher.

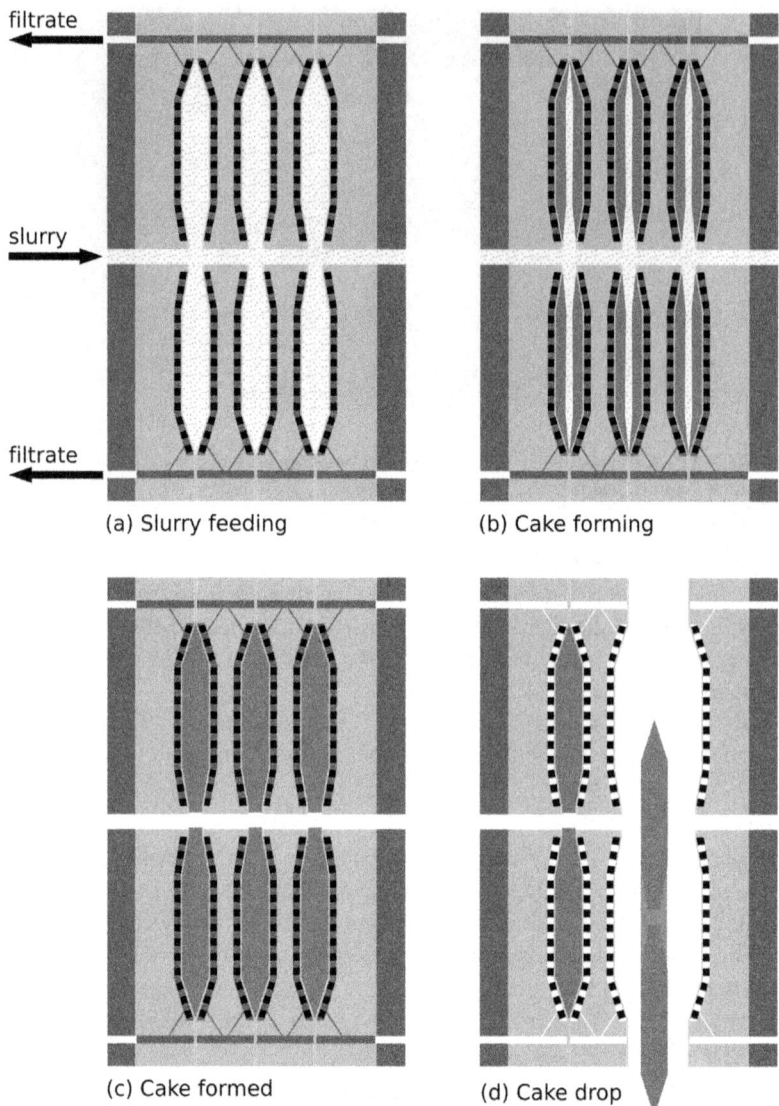

FIGURE 7.24 A simple filter press cycle.

one side of the cake to the other while the cake remains pressed, minimizing the chance of any disruption to the cake structure.

7.3.1.4 Air Drying

Once again, there are many possibilities here, but it is also possible to pass air through the cake while it is still being pressed.

7.3.1.5 Cake Discharge

Once the filtration process steps have finished, the plate pack will open, driven by the hydraulic cylinder.[9] The cakes are then removed, by, for example:

- a mechanism to shake the plates when they are open
- a mechanism to shake the cloths
- a mechanized scraper arrangement that moves from chamber to chamber and scrapes the cake from the cloth
- a person removing the cakes one at a time with a scraper, or even a hand.

Figure 7.24 shows a very simple cycle, from slurry feeding to cake discharge (omitting air drying).

Because of the large number of sealing surfaces formed by cloths, there may be a certain amount of liquid dripping from the filter. If this could cause an issue (it might make a rubber conveyor belt wet or could fall into dried cake), then some form of movable barrier is normally used. This might be a drip tray that can be moved by hand on a small filter, or an actuated bomb-bay door on larger filters.

Given that the cake discharge will be intermittent, the cake handling system must be designed to cope with large amounts of cake from time to time and will be more substantial (and expensive) than a cake handling system for a continuous filter with the same daily throughput.

7.3.1.6 Cloth Cleaning

Cloth washing can be done manually or using a mechanism that passes through the plate pack (see Figure 9.12). In either case, the cake chute should be covered to prevent this wash liquid from flooding through the cake handling system.

7.3.1.7 Applicability

Filter presses are used in a huge variety of different applications, from dewatering of mineral mining slurries to blood plasma purification. They are, perhaps, the most versatile and certainly the most widely used of all of the filters supplied in the past twenty years.

7.3.2 Tower Press

The tower press originated in the Ukraine in the 1950s, although it was a number of years before this type of filter became a widely adopted technology, in large part because of advances in cloth manufacture and control and instrumentation.

---

[9] The plate pack can open all at once or in smaller groups (perhaps one at a time). This depends upon the process requirements and the cycle (if the cycle time is eight hours, then twenty minutes to discharge cake is less of an issue than if the cycle time is ten minutes).

The tower press, in its basic form, is a filter press that has been turned so that the plates lie horizontally and so are arranged on top of each other in a vertical stack. In this case, the cloth acts as both the filter medium and the conveyor for cake discharge.

Some of the outcomes of this arrangement are that:

- the cake discharge time can be very short, since the cakes from each plate leave the filter at the same time
- the cakes are formed with gravity acting favorably and tend to be very uniform
- the cloth can be readily cleaned.

The frame construction still performs the same job as a filter press: to open and close the plate pack and keep it together despite the enormous forces trying to push it apart.[10]

There are several possible cloth arrangements, but three common ones are:

- Single cloth – endless: This is the arrangement shown in Figure 7.25. In this case, the cloth passes through the vertical stack of horizontal plates, passing over a roller at the end of each.
- Single cloth – end to end: Similar to the endless cloth arrangement, the cloth winds from one roller, through the plate pack onto another, like a cassette tape. The cloth advances a few plate-lengths per cake discharge cycle and once there is no more left on the first roller, it rewinds back from the second roller.
- One cloth per plate: Exactly as it says.

Figure 7.26 shows a tower press plate in more detail. The plates may incorporate a pressing diaphragm, if the process requires one.

7.3.2.1 Slurry Feeding

Typically, slurry is fed to each chamber through a flexible hose (via a common manifold), although it is also possible to form a slurry channel in the same way as in a filter press.

7.3.2.2 Cake Pressing, Cake Washing and Air Drying

As with filter presses, the cycle proceeds with pressing (optionally), cake washing (optionally) and air drying. The uniform cake that has usually formed means that the washing and drying are particularly effective.

---

[10] It could also be possible to keep the plate pack closed using a large weight; to prevent a 6 m^2 tower press from opening when it is operating at 16 bar, you would need a few fully loaded Airbus A380 airliners, with 500 passengers in each of them. This would probably introduce other practical difficulties, opening the plate pack for cake discharge being one.

Chapter | 7 Filter Design

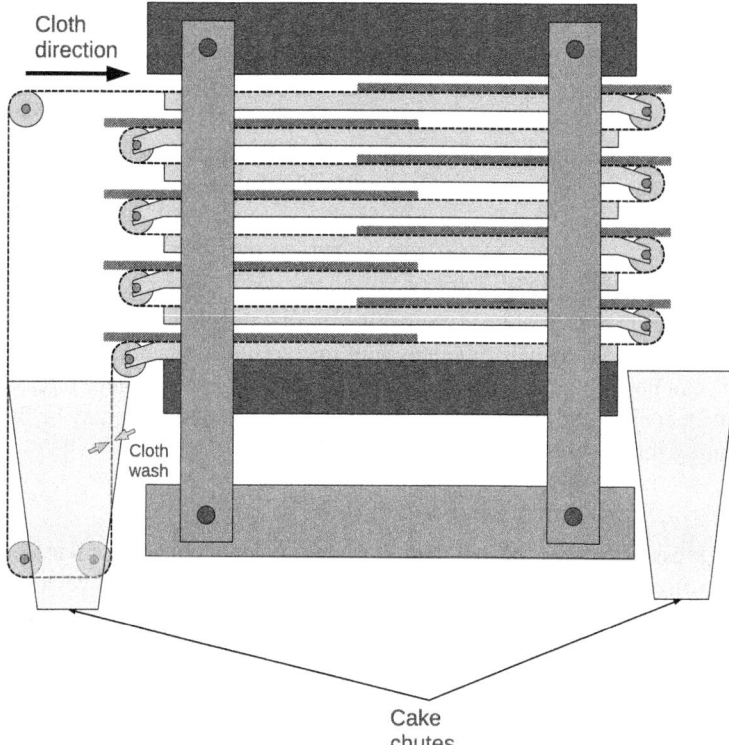

FIGURE 7.25 General arrangement of a single-cloth tower press, shown during cake discharge; cloth tensioning and cloth tracking omitted.

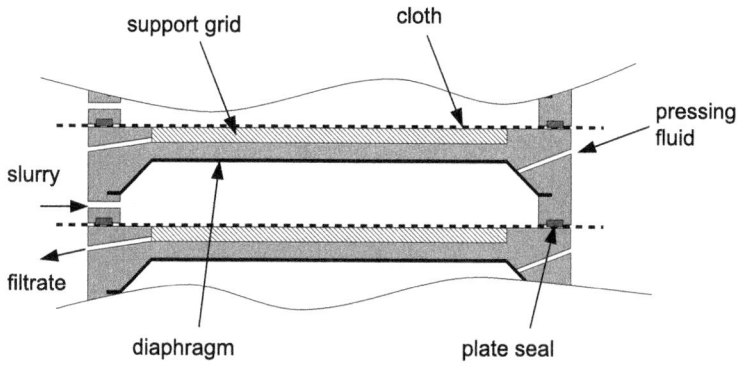

FIGURE 7.26 Tower press plate.

7.3.2.3 Cake Discharge

The plate pack opens and the filter cloth, or cloths, advance, so that cake is discharged. A small scraper may be needed on each plate to make sure that the discharge is complete. (Figure 7.27).

In certain circumstances, these filters can discharge several tonnes of filter cake in a few seconds, presenting a significant challenge to the cake handling system.

7.3.2.4 Cloth Cleaning

A particular advantage of this type of filter is that the cloth can be cleaned during cake discharge. Equally, the length of the cloth movement during cake discharge can be tailored so that alternative sides of a particular piece of cloth are used between cycles. Therefore, any solids pushed into the cloth may be flushed out during the next filtration cycle.

7.3.2.5 Options/Alternatives/Variations

Tower presses can also be supplied with double-sided plates, similar to those used in filter presses, so that filtration capacities per plate can be increased and alternative process cycles are available (e.g. pressing during cake washing through the filtrate channels).

Tower presses can also be readily expanded, provided the frame structure is large enough, with the addition of more plates.

7.3.2.6 Applicability

Automatic tower press filters are best known in the mining industry operating on base metal concentrates. Other key applications are starch and certain industrial minerals applications.

FIGURE 7.27 Cake discharge (Outotec Filters).

7.3.3 Tube Press

The tube press originated in Cornwall, UK, during the 1970s (within a few miles of the plant shown in Figure 2.1). The main reason for its development was to apply very large dewatering pressure to fine kaolin slurries to minimize the amount of water in the filter cake, sometimes eliminating the need for thermal evaporation.

Of all large-scale slurry filtration equipment available, the tube press delivers the greatest dewatering pressure, greater than 150 bar. Therefore, it tends to be used in very demanding applications with fine, compressible, materials.

The very high dewatering pressures are safely contained by the inherent strength of cylindrical construction. (This is why divers have small cylindrical air tanks on their backs and not massive slab-shaped tanks.)

The tube press comprises a cylindrical outer casing, which incorporates a cylindrical bladder, and an inner candle that is covered in fine cloth (over layers of backing cloth). The candle incorporates filtrate channels (Figure 7.28) and is movable (up and down).

Individual tube presses are limited in size, but large numbers of units can be installed together to make a plant with a very significant overall capacity (Figure 7.29).

Although each individual unit is not expandable, a tube press installation can be extended by adding more units. Since the pressure pumps, vacuum and air supplies come from a common system, extra units can be added (or removed) relatively easily, provided the system has enough capacity.

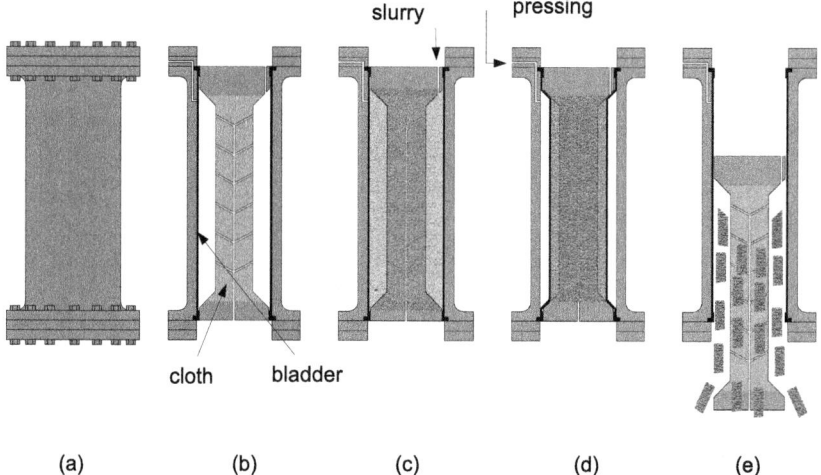

FIGURE 7.28 Tube press: (a) outer cylinder; (b) cross-section showing the central cloth-covered candle; (c) slurry feeding; (d) cake pressing; (e) cake discharge.

FIGURE 7.29 A tube press installation.

7.3.3.1 Slurry Feeding/Filtration

The slurry is delivered to the annular chamber formed by the cloth-covered candle and the outer cylinder. Normally, the chamber is just filled with slurry and there is not a prolonged filtration phase.

The bladder is then pressurized with hydraulic fluid (usually water) to extremely high pressure. In fact, at over 150 bar, this is an order of magnitude greater than any other large filtration device. Filtrate flows through the cloth and out through channels in the candle.

This forms an extremely compact cake on the outer surface of the candle.

7.3.3.2 Washing

Washing processes are possible with a tube press. After pressing, the bladder can be drawn back and the annular gap between the bladder and the cake filled with wash liquid. The bladder is then pressurized again and the wash liquid pushed through the cake.

7.3.3.3 Air Drying

In the same way as other chamber filters, air can be passed through the filter cake, either directly using pressure provided by a compressor or by filling the chamber with air and repeating the bladder press.

7.3.3.4 Cake Discharge

After pressing, the bladder is pulled back with vacuum and the chamber vented to atmosphere. The candle is dropped by a small amount and cake discharged (usually a small back-pulse of air through the cloth helps with this discharge).

7.3.3.5 Cloth Cleaning

The tube press does not normally incorporate a cloth washing system, but the candles can be lowered for washing.

7.3.3.6 Options/Alternatives/Variations

Other, smaller horizontal tube presses have been produced.

7.3.3.7 Applicability

Tube presses originated because of the need to apply very large dewatering pressures to kaolin and other very fine industrial minerals applications. These remain very important applications for tube presses, but other applications where exceptional filtrate clarity or low cake moistures are needed have also been found.

If the solid particles are particularly fragile and likely to be broken (undesirably) by large forces, then the tube press is probably not the technology for that particular application.

7.3.3.8 Installation Notes

Typically, a conveyor runs underneath a line of tube presses and directs the cake to a common point. Because of the modular nature of a tube press plant, any installations should take possible expansion into consideration.

7.3.3.9 Maintenance Notes

As with other chamber filters, care should be taken to ensure that the general area around the tube presses is clean. Given the extremely high pressures, the condition of seals and cloths should be carefully monitored.

7.3.4 Candle Filters

In a candle filter, the filter elements, usually bundles of perforated tubes sheathed with a filter cloth, are hung from a manifold mounted in a pressure vessel (Figure 7.30).[11] There are a large number of possible process combinations, for example multiple washes.

Candle filters can operate with a relatively simple control system, since there are no moving parts to track and the entire process amounts to opening and closing valves in the correct sequence, controlled by either timers or other parameters (e.g. depth, pressure, filtrate conductivity/turbidity).

---

[11] I worked on a project once where we installed a filter with four candles ....

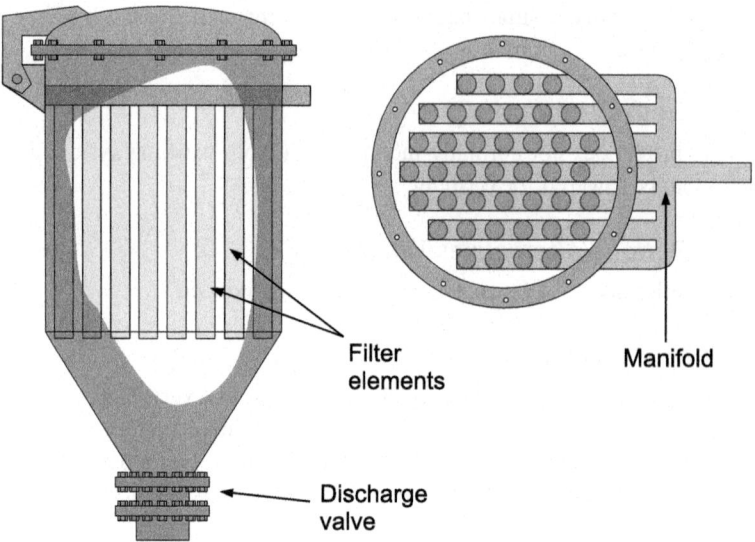

FIGURE 7.30 Candle filter: general arrangement.

7.3.4.1 Slurry Feeding/Filtration

The vessel is simply filled with slurry, usually with the pressure vessel vented to atmosphere, until a set point is reached. Once the vessel is full, the filtrate valves are opened and filtration commences as liquid flows through the cloths, via the candles and out through the filtrate manifold.

Filtration can proceed for some time, and a filter cake forms on the outer surface of each of the candles. As the filter cake thickness increases, the apparent filtration area of the filter actually increases quite rapidly.

7.3.4.2 Washing

Once a cake has formed, the slurry feed can simply be switched over to wash liquid, and a washing process begins. The remaining solids in suspension in the pressure vessel will join the filter cake.

Another possibility is to drain the unfiltered slurry in the vessel back to the feed tank, using air pressure to keep the cake in place on the candles and then filling the vessel with wash liquid.

Alternatively, a spray system can be used to minimize the amount of wash liquid used for washing, so that the vessel does not need to be filled with wash liquid. In this way, a mist of wash liquid passes through the filter cake.

Counter-current washing is possible, although it relies upon a system of tanks, pumps and valves.

7.3.4.3 Air Drying

Once the liquid level has dropped below the bottom of the filter elements, the small amount remaining (the heel) can be drained back to the feed tank.

Chapter | 7 Filter Design

Air drying can continue, either for a predetermined period or based upon a parameter (e.g. once air flow increases beyond a certain point).

7.3.4.4 Cake Discharge

Cake discharge can take a number different forms. However, at its most basic, after air drying, the vessel would be depressurized and the cake discharge valve (usually a knife-gate) in the base of the unit opened. A sharp back-pulse of air into the filter elements inflates the cloths and dislodges the cake, which falls out of the base of the unit (see Figure 7.31). This back-pulse can be applied to all of the candles at once, or to smaller groups of candles.

An alternative is a slurry discharge, in which the cake is discharged from the candles with the pressure vessel closed and filled with liquid.

It is even possible to discharge cake (as a slurry) from some of the candles while filtration is proceeding on the other elements.

7.3.4.5 Cloth Cleaning

In many applications, the action of cake discharge alone (back-pulse of gas) is enough to keep the cloths clean enough for the following cycles. It is also

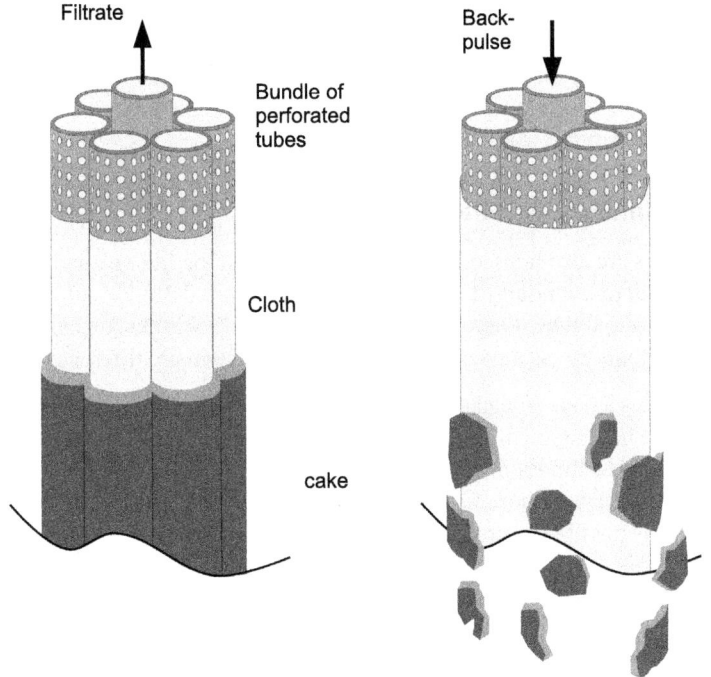

FIGURE 7.31 Schematic of a candle filter element. As cake forms on the cloth on the outer surface of the element, filtrate passes through the bundle of tubes. During cake discharge (right) a black-pulse of fluid causes the cloth to expand, dislodging the cake.

possible, with the appropriate combination of valves, to back-flush the cloths with liquid (even filtrate).

Given the enclosed construction of the filter, it is also possible to chemically wash the cloths (say in acid or caustic) to remove precipitated material from the cloths.

7.3.4.6 Installation

If the cake is to be reslurried then the candle filter can be positioned to discharge directly into an agitated tank. However, typically, candle filters are arranged along a belt conveyor.

7.3.4.7 Options/Alternatives/Variations

Despite the apparent simplicity of candle filters (or perhaps because of it), there is a large selection of alternatives to choose from in terms of materials of construction. Candle filters can be supplied with some manifold connections blanked off, so that other candle elements can be added later to expand the capacity.

Candle filters can be used as precoat filters.

7.3.4.8 Applicability

Candle filters have found many suitable applications, often as a result of their lack of moving parts, enclosed process and versatility.

Given that there are no moving parts or bearings inside the pressure vessel, they can be used in relatively harsh environments and the unobstructed route of the cake discharging from the pressure vessel.

7.3.5 Spinning Disc Filters

These filters are also housed in a pressure vessel, but the filter elements are a set of horizontal discs mounted on a central shaft (Figure 7.32).[12]

In operation they offer similar processing possibilities to candle filters (including their use as a precoat filter). However, the main differences are:

- The plates are horizontal and so the cakes will remain there, even if there is no pressure holding them in place.
- At the end of filtration, any slurry remaining in the vessel can be directed to a scavenger plate (whose filtrate line is closed during normal filtration). In this way, it is possible to produce a cake with no slurry wasted or returned to the slurry tank.
- Cake discharge occurs when the filter discs spin, throwing the cake outwards and down the inside of the pressure vessel.

---

[12] There are variations in which the shaft is horizontal and the discs vertical, but these are less common.

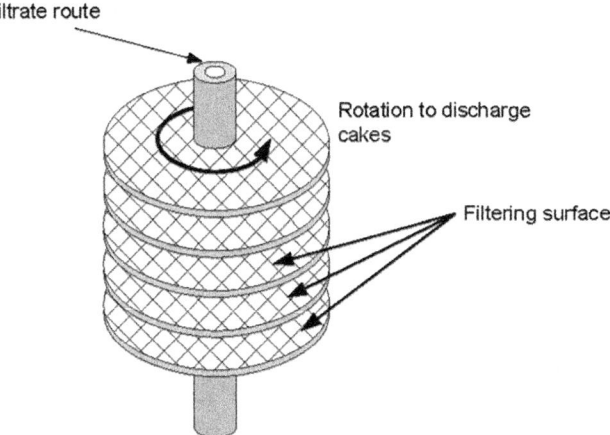

FIGURE 7.32 Disc stack in a centrifugal discharge filter.

- Spinning disc discharge filters introduce bearings into the pressure vessel and these may interrupt the route that the cake takes out of the vessel.

7.3.6 Leaf Filters

The final variation for batch pressure vessel filters is the pressure leaf filter, which resembles a candle filter (many of the parts are interchangeable) except that flat, cloth-covered leaves are used instead of candles. In other respects the operational cycle is similar.

7.4 CENTRIFUGAL FILTRATION

7.4.1 Batch Centrifuges

A batch centrifuge comprises a rotating basket, oriented horizontally or vertically. A simple schematic is shown in Figure 7.33.

7.4.1.1 Slurry Feeding/Filtration

The basket is accelerated so that the feed is spread uniformly over the filter cloth. Once the basket has been fully charged, the basket accelerates to full speed. This speed can be greater than 1000 rpm. While there is a continuous liquid phase throughout the basket, this provides a motive force for filtration, rather like a pressure difference. Once the liquid level drops below the surface of the cake, the deliquoring continues in which the centrifugal force (trying to throw droplets of moisture outwards) competes with surface tension within the cake's capillaries.

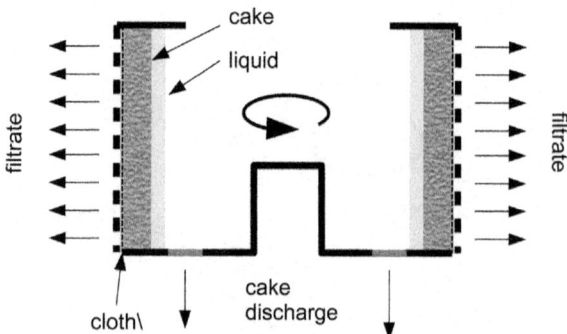

FIGURE 7.33 Schematic of a batch centrifuge. After the batch operations are completed, a paddle (not shown) moves in to direct the cake through the gap in the base of the basket.

7.4.1.2 Washing

Wash liquid can be applied to the basket (usually as a spray) so that the cake can be washed. Once the cake has been formed there is no practical limit to the number of washes that can be applied.

7.4.1.3 Gas Drying

Pressurized centrifugal filters can further enhance the filtration process and even provide a gas-drying stage to the filtration cycle, but this is not common.

7.4.1.4 Cake Discharge

After the cycle is complete, different methods can be used to remove the cake from batch centrifuges:

- Removal by scraper or paddle: A paddle or plough moves into the cake and directs the cake out of the basket (which is rotating slowly). If the particles are particularly fragile, this scraping action may damage them. Since the scraper does not usually scrape against the cloth itself, there will be some residual cake left in the basket (also known as the heel).
- Removing or inverting the cloth: In this case, the entire filtering cloth is either removed completely (feasible if the product is high value) or turned inside out. Since there is no scraping action, this means that there will be relatively little damage to fragile particles. There will be no heel left in the basket after discharge.

7.4.1.5 Options/Alternatives/Variations

Filtering centrifuges are often used in food and pharmaceutical production processes that must comply with good manufacturing practice. Compared with general industrial filtration equipment, some centrifuges resemble sculptures, with polished smooth surfaces with no hold-up areas or nooks.

A system of spray bars and balls can be used to clean the equipment completely.

7.4.1.6 Applicability

Filtering centrifuges can be applied in many applications, from general dewatering through food applications (starches) to the washing and isolation of pharmaceutical products and intermediates.

7.4.1.7 Installation Notes

Given that a large mass is rotating at very high speed, substantial foundations are needed to keep the units tethered to the floor and to isolate the building from any vibration.

7.4.1.8 Maintenance Notes

Very high-speed rotating equipment has its own set of standards, but these will include regular checks on the integrity of the mechanical components and, in particular, inspections for signs of fatigue.

Any imbalance in a large mass rotating at very high speed creates the potential for a very significant incident.

7.4.2 Continuous Centrifuges

Continuous centrifuges use the same basic principle, in which centrifugal forces provide the motive force for filtration. However, the basket is fed continuously with slurry and discharges cake continuously.

The same general points regarding installation and maintenance apply. It is important to note that the cake is continuously disrupted in continuous centrifuges, so filtrate may not be as clear as in an equivalent batch centrifuge.

7.4.2.1 Cake Discharge

Two main forms of continuous centrifuge are:

- Conical basket: In this case, a conical basket (at an appropriate angle) is fed continuously with slurry. The centrifugal action provides both the motive force for filtration and moves the dewatered (and, if applicable, washed) cake towards the outer edge of the basket.
- Pusher: A reciprocating plate moves in and out of the basket, pushing cake out (see Figure 7.34). (Think of a penny-push machine in an amusement arcade.) Pusher centrifuges can have more than one stage, with a set of concentric baskets and pusher plates.

7.4.2.2 Applicability

Continuous centrifuges tend to be used for relatively free-filtering materials, such as coarse industrial minerals and salts.

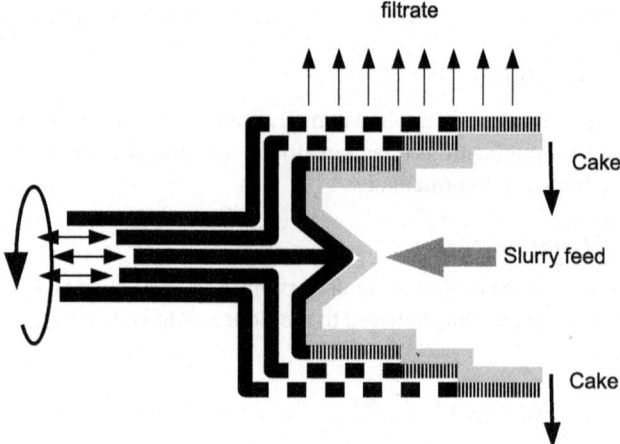

FIGURE 7.34 Schematic of the rotating parts in a pusher centrifuge-spinning bowl with at least one inner bowl that rotates and moves backwards and forwards, pushing cake towards a cake discharge region. Given that this type of machine is normally used for very coarse solids, the cake is often dry and powdery.

7.5 SUMMARY

There are many filtration types. In this chapter they have been categorized according to the mode of operation (continuous or batch) and the driving force for filtration used (vacuum, pressure or centrifugal).

Within each filter type there exist many possible variations, options and alternatives. Approaches to find the best filter for a particular duty are discussed in later chapters.

Chapter 8

Filter Installation

Chapter Outline
8.1 Human Considerations	126
8.1.1 Access: Walkways, Platforms, etc.	126
8.1.2 Noise	126
8.1.3 Lighting	126
8.1.4 Ventilation and Extraction	127
8.1.5 Flooring	127
8.2 Process Considerations	128
8.3 Slurry Systems	129
8.4 Cake Handling	129
8.5 Summary	129

The way that a filter is installed can play a critical part in the success or otherwise of the process and, as with all of the other success factors, a poor installation can harm the overall success of the process, even if everything else is in good order. Care should be taken to ensure that the particular needs of a filtration process are taken into consideration in a plant layout. This chapter will look at this in two ways, from the point of view of:

- the people who will operate and maintain the filter
- the process (the slurries, cake, water, filtrate and air that flow throughout it).

When considered against the other success factors, in terms of difficulty and timing, major changes to a filter installation may take several weeks or months and would be relatively involved, requiring a significant amount of project work.

This chapter will deal with some of the main issues; however, most filter equipment suppliers can offer installation advice and it is always useful to visit similar installations, where possible, to learn about the good ideas – and the pitfalls. A company will follow its own internal and statutory standards for machinery and plant installation, but there may be scope to pay particular attention to some of the issues outlined below.

A review and project to modify a filter installation can save a filtration process. It is useful to hold a review of these issues, say one month after a filter has been commissioned.

8.1 HUMAN CONSIDERATIONS

The vast majority of filtration processes need daily, and sometimes continuous, attention from operators. If paying such attention is unpleasant, then the filtration performance will ultimately suffer. The success of the process can be improved in a number of ways.

8.1.1 Access: Walkways, Platforms, etc.

Many of the filter types discussed in Chapter 7 are open, so that you can see the process taking place, in particular the continuous vacuum filters: horizontal belt, pan, rotary drum and rotary disc. Seeing the process means that small problems – with slurry or wash liquid distribution, uneven cake, cake discharge or cake cracking – can be spotted quickly and easily. It makes sense, then, to give the operators convenient and safe access to make the most of the advantage that open continuous processing gives.

In addition, the needs of maintenance staff should be taken into consideration so that inspection and spare part replacement can be as simple as possible. One extra platform that gives access to a group of critical components could make all the difference between timely replacement and an unexpected critical failure.

8.1.2 Noise

None of the filters described here will operate in silence, especially if something has gone wrong and, say, air is whistling through a torn cloth under vacuum. However, unnecessary levels of noise will lead to:

- fewer visits to the filter
- shorter visits to the filter
- poorer communication during visits to the filter (it will be harder to hear other people speak).

It makes sense to build good noise suppression into the installation, for example mufflers for any venting of air.

If the process is still particularly noisy, then a sound-proofed quiet room nearby could not only give people a break from the noise, but also give them somewhere to remove their ear defenders and discuss anything unusual that they have noticed, as well as any ideas that they have for improvements.

8.1.3 Lighting

Filtration equipment can be quite intricate and small problems can amplify themselves if they are not dealt with promptly. For example, a piece of filter cake can lodge in the wrong place and, over a period of a few days, or even hours, can lead to significant problems.

Given that filtration plants tend to be quite dusty and humid, lamps can become dirty and lighting can quickly become inadequate, even if, when clean, the lamps offer enough light. Making as much use as possible of natural light and overspecifying the lighting arrangement slightly (to allow for fogging of the lights) will pay you back. Given that lamps are likely to be come dirty, it makes sense to design a lighting arrangement that can be cleaned from time to time.

8.1.4 Ventilation and Extraction

Filtration plants can also be unpleasant because of steam and dust. An extraction hood over filtration equipment (especially continuous vacuum filters with hot wash liquid) will make a more pleasant working environment where more will get done.

It will also reduce the amount of waterborne dust that settles on critical components.

8.1.5 Flooring

Filtration processes can often be surrounded by pools of liquid, either from the process itself or from hosing down and cleaning.

A puddle of liquid on the floor near a vital component will mean that this component does not receive the attention that it could; the same applies to a critical point for viewing the process.

Ideally, a solid concrete floor with adequate sloping towards drains or a mesh floor over a solid floor with drainage slopes will make it easier to view the process and inspect and change parts.

Figure 8.1 shows an example of a good installation. The area is well lit and there is good access around the filters for maintenance. The control panels are

FIGURE 8.1 An example of an installation (FLSmidth).

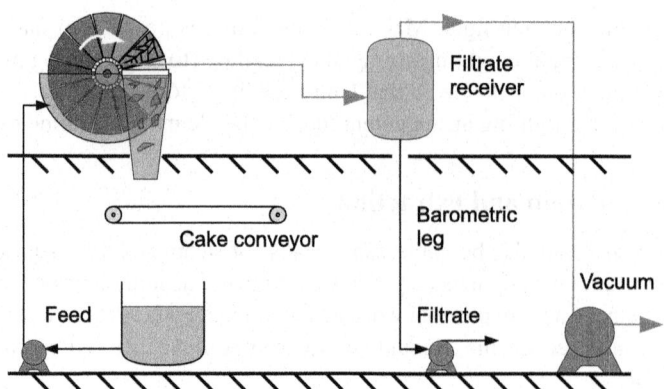

FIGURE 8.2 An example of a simple vacuum filter (continuous) installation.

easily accessible and the area can be kept clean. There is also a crane above the plant so that heavy items can be moved easily during major overhauls.

8.2 PROCESS CONSIDERATIONS

The installation should allow all process flows, i.e.

- slurry
- filtrate
- air (under pressure or vacuum)
- cake
- wash liquids/wash filtrates,

to pass freely without let or hindrance. It is surprising how the performance of many filtration plants are hindered by limitation to one or more of the above.[1]

Pumps for slurry, air (or vacuum), filtrate and other utilities need to be sized for their appropriate duties, and an additional sizing buffer added to deal with surges in the process.

Figure 8.2 shows a typical installation for a continuous vacuum disc filter. The filtrate leg (a vertical drop) can be designed to have a barometric effect to enhance the performance; if this is the case then the elevation of the filter is important.

While this installation is for a continuous filter operation, batch filters require storage of slurry, filtrate and, usually, drying air.

In addition, if the filter itself is expandable, either all pumps and conveyors should have the capacity to deal with the extra throughput or it should be

---

[1] I have seen one installation where the user followed the equipment supplier's specification for a filtrate drain, a 4 inch pipe at the appropriate slope, except for a few meters of 2 inch pipe passing through a wall .... The filtration plant was not able to deliver full capacity until this was rectified.

possible to add to them to increase this capacity. Leave space on the floor for extra pumps if this is the case.

8.3 SLURRY SYSTEMS

Pumping slurry can be problematic, particularly if the solids settle rapidly. If this is the case, then the pipe layout and pumping arrangements must take this into account, usually with continuous pumping. (For batch filter systems the slurry would be recirculated to the feed tank.)

As discussed in Chapter 6, slurry pumping (and flow control) can affect the nature of the slurry by shearing particles or agglomerates of particles. Some low shear pumping options are available, for example progressive cavity pumps or even hose pumps. The type of valve can also be critical, see Section 6.3.3.

8.4 CAKE HANDLING

Filter cake can be an extremely difficult material to convey and it can be particularly troublesome if the cake properties vary. A belt may slide underneath a batch of wetter than expected cake, or a screw conveyor may burrow a tunnel through a pile of stationary cake.

There may be some advantage in giving the conveyor belts a slight backward slope (so that cake is conveyed upwards) to allow any liquid to drain back off the belt.

Where a filter is expandable, it makes sense to install a cake conveying (and slurry supply) system that will be able to cope with the larger throughputs.

8.5 SUMMARY

Designing a filter installation should take into account the needs of both the process materials and utilities flowing through it and the people who will operate and maintain the equipment.

Improving the working environment is relatively easy to do and will certainly bring benefit. A filtration process that can be seen and a filter that can be inspected and maintained easily will perform better. Installing observation platforms, improving lighting and muffling unnecessary noise will all help. The benefits will be seen particularly in productivity (because of higher uptime) and health and safety (less chance of strains or accidents).

Advice from equipment vendors (they have no interest in supplying a fantastic piece of equipment that will fail because of poor installation) and other users should be sought throughout filtration plant design.

Chapter 9

Filter Cloth

Chapter Outline

9.1 Desired Outcomes	133		9.2.4.1 Calendering	138
9.2 Cloth Design and			9.2.4.2 Coating	140
Manufacture	133		9.2.4.3 Heat Setting	
9.2.1 Materials	134		/Prestretching	141
9.2.2 Yarn Types	135		9.2.5 Cloth Finishing and	
9.2.3 Weave Types	136		Fabrication	141
9.2.3.1 Plain Weave	136	**9.3 Cloth Support Grid**		**141**
9.2.3.2 Twill Weave	137	**9.4 Cloth in Operation**		**142**
9.2.3.3 Satin Weave	138		9.4.1 Cloth Failure	143
9.2.3.4 Multilayer			9.4.2 Cloth Cleaning	144
Weaves	138		9.4.3 Cloth Repair	144
9.2.4 Postweaving Treatment	138	**9.5 Summary**		**146**

Now. Some fool has invented an indestructible cloth. Where is he? How much does he want?

Sir John Kierlaw, a cloth factory owner in the film *The Man in the White Suit* (1951)

The filter cloth is the most important single component on a filter; in fact, it may sometimes be referred to as simply "the filter".[1] It is the location of all of the action in the early part of any filtration cycle and it can play a key part in the success of a process. Many filtration processes have failed because of poor cloth selection or a lack of care for the cloth in operation.

Besides being the most important single component, it is also the most vulnerable part of most filter installations: a stray piece of welding splatter, a rough edge or a rogue sharp particle can spell the end of the cloth and then, usually, severely compromise the success of the process.

The title of this chapter, "Filter Cloth", already excludes many forms of filter media, including membranes, perforated or porous material, sintered metal media and ceramic media. This section will mainly be limited to cloths woven

---

[1] Remember the close link between the words "filter" and "felt".

Solid-Liquid Filtration. DOI: 10.1016/B978-0-08-097114-8.00009-5
Copyright © 2012 Elsevier Ltd. All rights reserved.

from polymer yarns. Non-woven cloths (felts, papers) and woven metallic cloths will not be discussed in detail.[2]

There are many suppliers of filter cloths and, over the past few decades, a great deal of development has taken place. Therefore, there are a large number of alternatives from which to choose: which type of cloth do you select, from which supplier and which of their specific products? Finding the most suitable cloth is often, as with many things to do with filtration, a matter of informed trial and error. Cloth suppliers and other filter users will have a wealth of experience, much of which could be relevant to a particular filtration duty. It is well worth asking for their opinion, since finding a suitable cloth can transform the overall success of a manufacturing process.[3]

The technology used in the manufacture of these cloths can be very impressive and the quality checks very thorough. Any weaving fault, no matter how small, will provide a vulnerability that will be sought out by abrasive particles in a slurry. The result may, in time, be a pinhole – at best leading to an increase in filtrate solids, at worst leading to a failure of the whole cloth and shutting down of the filtration process and, therefore, possibly, the entire process.

In some filter equipment types, the cloth may also be called upon to perform some other mechanical duty in addition to the task of providing a barrier to solids and a support for the filter cake. In some filters, the cloth acts as a conveyor or is given a back-pulse to remove (or help to remove) the filter cake, or it may be needed to seal against pressure or vacuum. We shall need to look at how cloths are produced to deliver capability to perform these additional duties.

At the same time as being the most important component on a filter, the cloth will often be the most costly spare part over the lifetime of a filter installation, so steps should be taken to ensure that:

- the correct selection is made
- the cloth is protected
- the cloth is used in a way that optimizes the performance and lifetime
- the cloth is replaced (or cleaned) at the appropriate time.

Filter cloths, unfortunately for process plants, have a finite life. At the end of this life, there is usually a reduction in performance and a failed cloth (holed, torn or worn out) may jeopardize the entire process or, worse, even the integrity of the filter machinery itself: a jet of abrasive slurry shooting through a holed filter cloth can burn a hole in polypropylene plates or even stainless steel. If there is an overriding aim for this chapter, it is to provide background information and

---

[2] A number of woven metal filter cloths are also available, as well as perforated metal sheets. However, since these are (for now) not often used on the filter types within the scope of this book, they will not be discussed further here. However, these metallic media could transform a number of processes and this will be a field of even more development in the future.

[3] Many equipment selection processes start with the filter design and not the filter cloth and optimal operating conditions. This seems to be the wrong way around.

ideas that will allow maximization of the productivity, lifetime and utility of filter cloths; in other words, how to get the most out of them.

In terms of involvement and difficulty, a small project to investigate a new filter cloth material could be completed in only a few weeks, including discussions with vendors, small-scale testing and a plant trial.

9.1 DESIRED OUTCOMES

When considering cloth selection, it is important to have a clear idea of what is required from that cloth, in the terms set out in Chapters 4 and 5. You need to know how much you value an improvement in cloth performance. For example, how much can you increase productivity if you improve filtrate clarity? How much is each extra kilogram, or tonne, worth?

Now, you may find that you need to sacrifice performance in one area to achieve an improvement elsewhere. For example, if you select a cloth that improves the clarity of the filtrate, it is likely (although not certain) that this would be at the cost of a reduction in throughput. Equally, there may be a platinum-impregnated cloth that gives the best overall filtration performance, but which costs 100 times more than a normal cloth that performs almost as well.

A useful checklist should include a judgment on:

- filtration process performance
  - filtrate clarity
  - throughput
  - cake moisture
  - washing result
  - overall cost of the process (including, but not limited to, cloth costs)
- cake discharge performance
- consistent performance
  - cleanability
  - resistance to blinding
- durability/lifetime (and hence cost per unit of production).

It is possible to gauge the filtration process performance of the basic cloth from a laboratory test, but assessment of the consistency of performance (cleanability, resistance to blinding) and lifetime can only come from experience of similar processes and/or an extended pilot- or even production-scale trial.

9.2 CLOTH DESIGN AND MANUFACTURE

The term "design" in the title of this section is used deliberately and is appropriate since many large filter-cloth suppliers are no longer companies that simply have a number of weaving machines and a salesforce. Filter cloths are more and more being deliberately engineered for their final duty and can incorporate all manner of innovations in their construction.

In turn, this section will look at the materials of construction, yarn types, weave types of the basic fabrics, then at the postweaving treatments and, finally, cloth finishing and fabrication or how the cloths are made suitable for the intended filter machinery. The aim is to provide a background to the terminology and how it may relate to a particular process.

9.2.1 Materials

The majority of woven cloths used on solid–liquid filters of the type discussed here are made by weaving polymer yarns. The three most common polymers are polyester, polypropylene and polyamide (nylon). There are other, usually more exotic and expensive alternatives, for example polyvinylidene fluoride (PVDF), polytetrafluoroethylene (PTFE) or even natural fibers, but these account for a small proportion of the cloths in use.

There can be major variations within even these broad names for the materials (just as there are major differences between the various grades of stainless steel) and there will be a variation in quality of what should be the same material from different suppliers. However, it is useful to tabulate some of their general features and applicability (Table 9.1).

TABLE 9.1 General Properties of Some Filter Cloth Materials

	Polyester	Polypropylene	Polyamide
Buoyancy in water	Sinks	Floats	Neutral
Water absorption	Low	Very low	Check
Resistance to temperature (>80°C)	Good	Check	Good
Resistance to abrasion	Good	Check	Very good
Resistance to alkali	Check	Very good	Good
Resistance to mineral acid	Good	Very good	Check
Resistance to organic acid	Check	Very good	Check
Resistance to oxidizing agents	Good	Very good	Check
Resistance to organic solvents	Good	Check	Good
Resistance to sunlight	Good	Poor	Poor

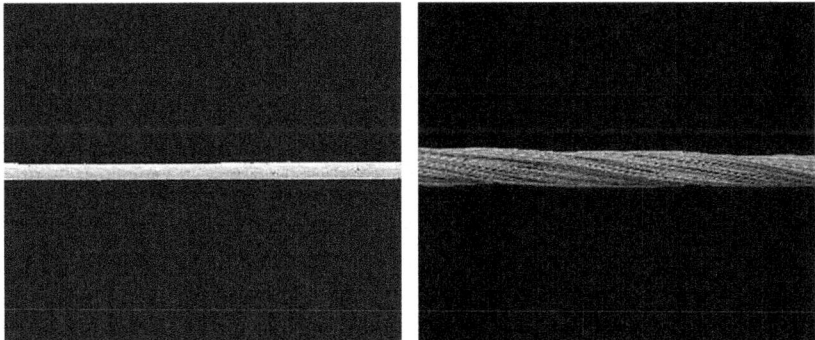

FIGURE 9.1 Left to right: Monofilament and multifilament yarns (Clear Edge Ltd).

The reason for including the buoyancy in water in Table 9.1 is that it offers a relatively simple way of identifying a material. A piece of cloth is saturated in water and then put into a bucket: if it floats it will probably be polypropylene, if it sinks then probably polyester, if it sinks slowly then it could be polyamide.[4]

This table is not meant to be the final word in cloth suitability, but rather to act as a tool to help in the first cut when selecting a filter cloth. For a high-temperature, very abrasive slurry, polyamide will probably be the best material to evaluate first. This is not to say that for other, more subtle reasons, polypropylene would not be suitable.

Filter cloth suppliers have information on the use of their products in all manner of applications, so they should be the next port of call.

9.2.2 Yarn Types

The two main yarn types found in filter cloths are monofilament and multifilament. There are others, but the vast bulk of cloths will be one of these two (Figure 9.1).[5]

A monofilament yarn is made from a single piece of polymer, drawn out like a fishing line, and it resembles a long human hair. No liquid can flow through the yarn itself, only through the pores in the woven cloth. In operation, a monofilament cloth may be vulnerable to chemical attack and can become brittle. On the positive side, monofilament yarns are highly consistent and can be used to make extremely tight filter cloths.

Multifilament yarns are simply composed of a number of individual yarns that have been bundled or, sometimes, wound together. Liquid may pass through the yarn itself in some cases. This type of yarn may also offer benefits in terms

---

[4] If it sinks like a stone then it may well be something more exotic like PTFE.
[5] Alternative yarn types include spun yarns (short fibers spun together to make longer yarns) or tape yarns (flat fibers).

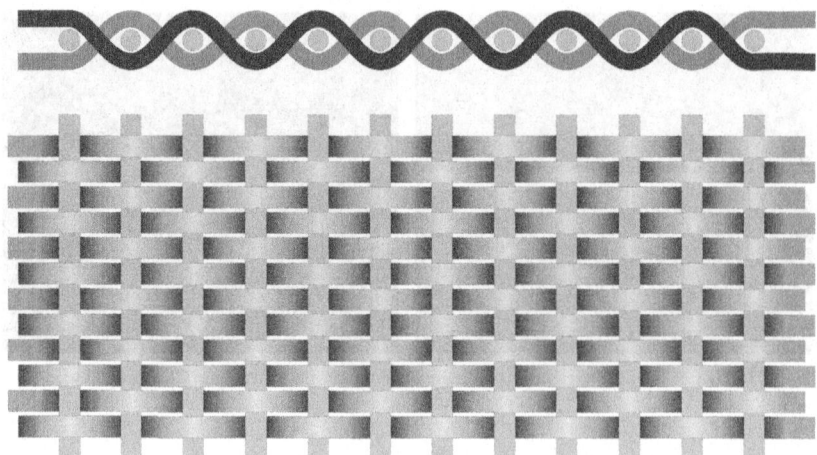

FIGURE 9.2 Plain weave: the warp yarns are vertical and the weft go from right to left.

of the mechanical properties of the cloth, for example in terms of flexibility or resistance to abrasion or tearing.

Cloths can be manufactured with a combination of yarn types; for example, a belt can be made with both multifilament warp yarns (for strength or stiffness) and monofilament yarns (to deliver a specified filtration performance).[6]

9.2.3 Weave Types

The basic construction of a woven cloth has not changed a great deal since it was first invented, so ancient humans, when passing beer through a piece of cloth to remove the solids, probably used something which, on a small scale and to the casual eye, looked remarkably like the cloth on a modern filter.

A woven cloth consists of a series of weft yarns passed over and under warp yarns.

9.2.3.1 Plain Weave

In the simplest type of weave, the plain weave, the warp yarns pass over and under alternate weft yarns (Figure 9.2). This cloth will tend to be stiffer and more elastic than other cloths formed from different weaves using the same yarns. In many cases, this construction provides the best particle-retention properties and is commonly used where the cloth does not have a secondary duty to perform (e.g. cake conveying).

---

[6] The cloth shown later in this chapter, in Figure 9.13, has multifilament weft yarns together with monofilament warp yarns.

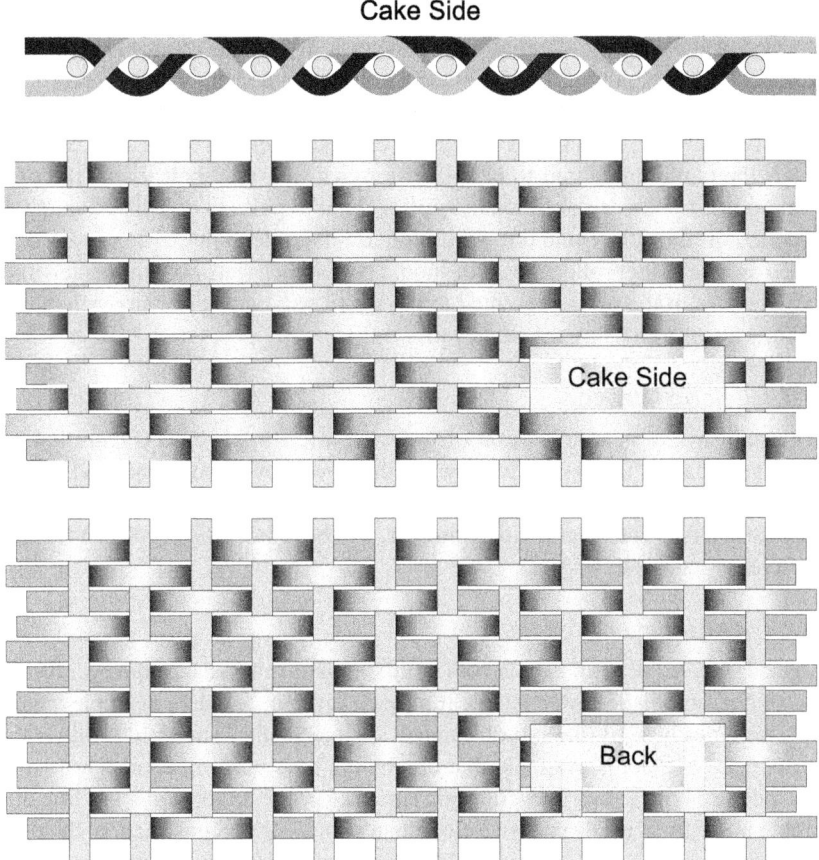

FIGURE 9.3 A 2/1 twill weave (over 2, under 1).

The plain weave also looks the same on both sides. So, a plain-weave cloth will have identical filtration and cake-release properties, regardless of which side is being used for filtration.

9.2.3.2 Twill Weave

The twill weave is a slight elaboration on the theme of the plain weave. In the example shown in Figure 9.3, the warp yarn passes over two weft yarns. Then the next warp yarn is stepped down. This type of weave produces a diagonal pattern in the cloth.[7] There are also some key differences that will affect filtration performance.

---

[7] Denim is probably the best known form of a twill weave. If you are wearing jeans, you will be able to see the diagonal pattern.

Unlike a plain weave, the cloth is no longer the same on both sides. The upper surface of the cloth (looking at the side shown in the illustration) is smoother; there is less weft yarn exposed to the cake than with a plain weave. Equally, the other side of the cloth is rougher. This excludes this type of cloth from some filter types, where both sides of the filter cloth are required to perform equally.

If the cloth is required to pass over a roller (e.g. on a belt filter), then the pattern in a twill weave will tend to give a slight impulse in one direction. After a period of time, this could throw the cloth off the roller. In order to remain centered on a belt-filter roller, the cloth should be symmetrical about the center-line.

The size of the pores presented to the slurry will be slightly larger than in a plain woven cloth using the same yarns.

9.2.3.3 Satin Weave

The satin weave is an extension of the twill idea. In this case (Figure 9.4), a weft yarn passes under one warp yarn, then over four, under one, and so on. The outcome of this is a very smooth upper surface and usually, therefore, good cake discharge. The reverse side, however, will be rather rough (you can feel the difference with your fingers).

9.2.3.4 Multilayer Weaves

In cases where the cloth acts as both a filter medium and a cake conveyor (e.g. a belt filter, automatic tower press), it may be difficult to select a simple weave that provides both filtration performance and mechanical strength. If this is the case, there are cloths available that combine a base made of coarser yarns, to provide strength, with a finer filtration surface (on one or both sides). While this might make for a more expensive cloth than a single-layer construction, the benefits of longer life while retaining filtration performance could make it a good choice. Figure 9.5 shows an artist's impression of a multilayer cloth, simply to give the idea (a real cloth may look nothing like this).

9.2.4 Postweaving Treatment

The woven filter fabric may require further treatment to improve particle retention, cake release or mechanical/thermal stability. Between the loom and fitting onto a filter, there are also several possible treatments that can be applied to the cloth to make it more suitable for its duty. The aim might be to improve lifetime or performance.

9.2.4.1 Calendering

Calendering is simply pressing the cloth with a very hot surface to soften, then flatten the yarns. This has two main effects upon the filter cloth:

- reducing pore size
- making the surface smoother, helping cake discharge.

Chapter | 9 Filter Cloth

This can have the effect of melting the filaments on the surface of a multifilament cloth together so that they appear and behave rather like monofilament yarns (Figure 9.6). Depending upon the end duty, the manufacturer may apply a number of calendering treatments, at a number of different temperatures.

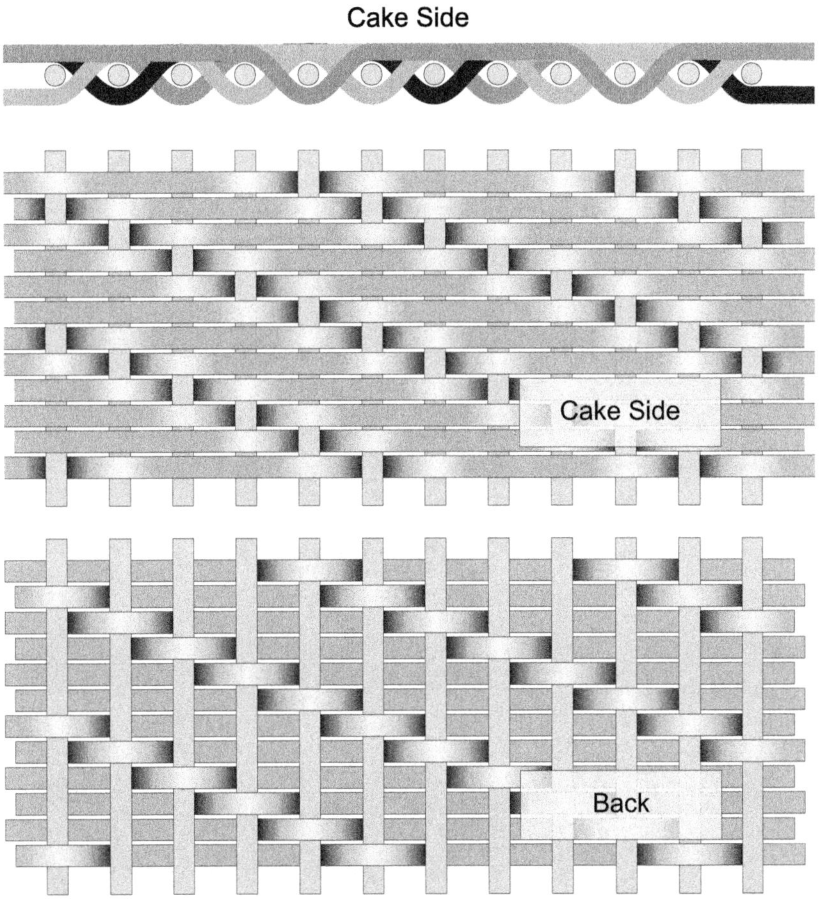

FIGURE 9.4 Satin weave (4:1).

FIGURE 9.5 Multilayer cloth, single side (cake on top).

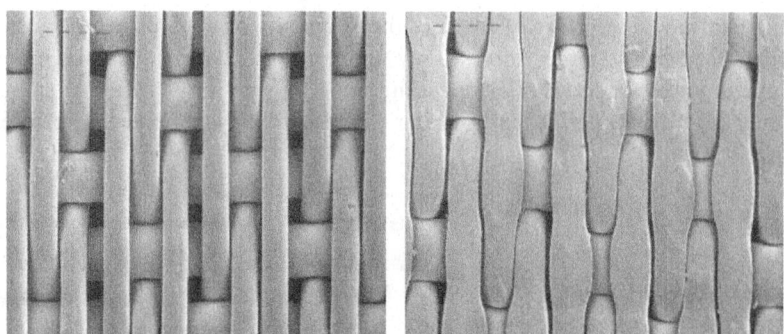

FIGURE 9.6 A twill-woven cloth before and after calendering (Clear Edge Ltd).

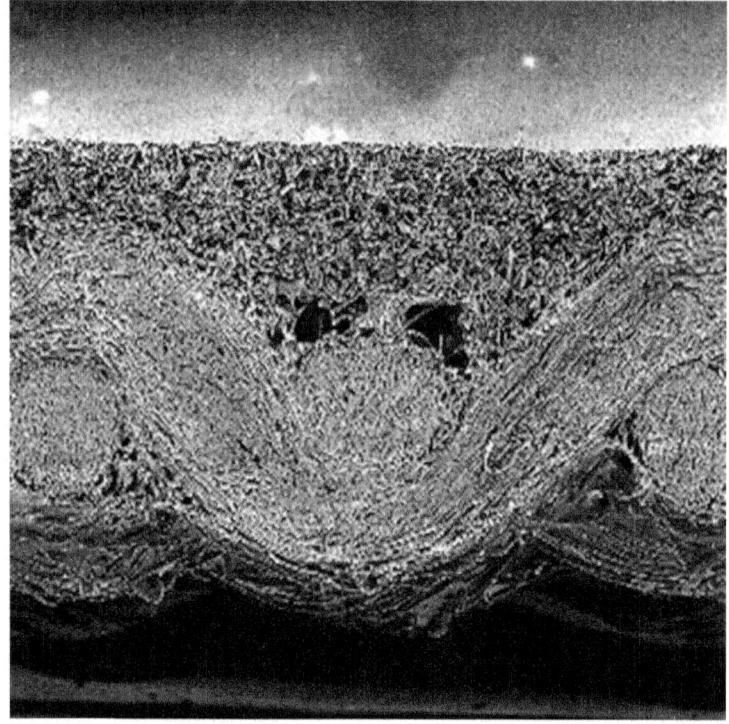

FIGURE 9.7 Coated cloth (Clear Edge Ltd).

9.2.4.2 Coating

Another way to improve the performance is to apply a coating material onto the woven cloth. This can deliver a marked improvement in terms of filtrate clarity and resistance to abrasion. Figure 9.7 shows a multifilament/multifilament plain woven cloth with a porous coating.

9.2.4.3 Heat Setting/Prestretching

Where the cloth is likely to operate under heat or tension (say on a belt filter) in the final application, it may suffer from thermal expansion (or shrinkage). This can affect the tension on a belt filter or the fit of the cloth into a filter press. For this reason, the cloth may be prestretched or preheated after weaving.

9.2.5 Cloth Finishing and Fabrication

The final fabric may still need to go through several steps before it can be used. It must be cut to the correct size and, perhaps, joined to make it suitable for the duty or to make it fit the geometry of the filter machinery.

This joining is usually performed by either stitching or welding and the choice of stitching thread may be overlooked. In some cases, a perfectly suitable cloth (in terms of material, yarn type and weave) can be let down by the wrong choice of stitching thread.

In addition, and depending upon the type of filter, it may be necessary to fit rubber seals or metal components into place to contain the dewatering force.

9.3 CLOTH SUPPORT GRID

All filters equipment must allow the filtrate to flow with as little restriction as possible. In some filters, with a very low filtrate flow rate, the cloth can lie directly on a flat steel surface, or possibly on a coarse backing cloth. However, in the majority of cases, the cloth lies on top of a grid (Figure 9.8) or possibly a series of molded pips (see Figures 7.22 and 7.23). The purpose of these is to provide support for the cloth and allow free passage for the filtrate. There are some possible issues to be aware of:

- Since support grids and pips are usually molded, sharp edges or burrs may be left after they have been manufactured. Poor finishing can also produce sharp edges (Figure 9.9).

FIGURE 9.8 The cloth support grid. Left: cloth side; right: underside, which allows the filtrate to flow freely.

FIGURE 9.9 A small burr (the injection-molding point).

- If the grid spacing is too large, the cloth may not be adequately supported and will stretch into the gaps, possibly damaging the yarns in the cloth or giving a focal point for cake cracking.
- If support grids are located in metal trays (e.g. in belt filters), they may have a different rate of thermal expansion and could either bow upwards or shrink to leave a long unsupported span for the cloth around the edge of the tray.

9.4 CLOTH IN OPERATION

Once the cloth has been selected and fitted, some steps can be taken to increase its lifetime. Given that cloths are often the most costly spare part and also the most critical, a clear strategy should be in place to protect both the success of the process and the machinery itself.

As discussed in Chapter 3 (and to be discussed in more depth in Chapter 11), the filter cake acts to protect the cloth from the damage and wear that come from particles passing through. The cloth is at its most vulnerable just as slurry reports to a clean cloth, when the filtrate (and hence particle) speed is the highest and there is no layer of pre-existing filter cake to provide a protective barrier.

9.4.1 Cloth Failure

A cloth has reached the end of its useful life when replacement is the best thing to do, in terms of the four dimensions of the success of a process. The mode of this failure is likely to be one of the following:

- Disintegration: The cloth starts to tear or come apart at the seams, perhaps because of chemical attack or wearing.
- Pinholing: Holes appear in the cloth, causing the filtrate clarity to be lower than the required level. The pinhole shown in Figure 9.10 was barely visible, but had a dramatic effect. Over time, this hole would have grown rapidly.
- Blinding: The capacity drops below the required level and cleaning will no longer refresh the cloth's performance.

There are also ways in which a failed cloth can lead to damage of the machinery, since most filtrate systems cannot handle slurry for a prolonged period. Where the cloth acts as a cake conveyor, a failure will cause a build-up of cake that can also damage machinery.

It should also be borne in mind that filter cloths do not fail in isolation; rather, it is the system comprising the cloth and the support grid that fails. One avenue for increasing lifetime is to look at the filter cloth support grid. Any rough edges should be eliminated. In addition, it may be possible to select a finer grid, so that the cloth has to bridge less of a span when it is subjected to the dewatering force.

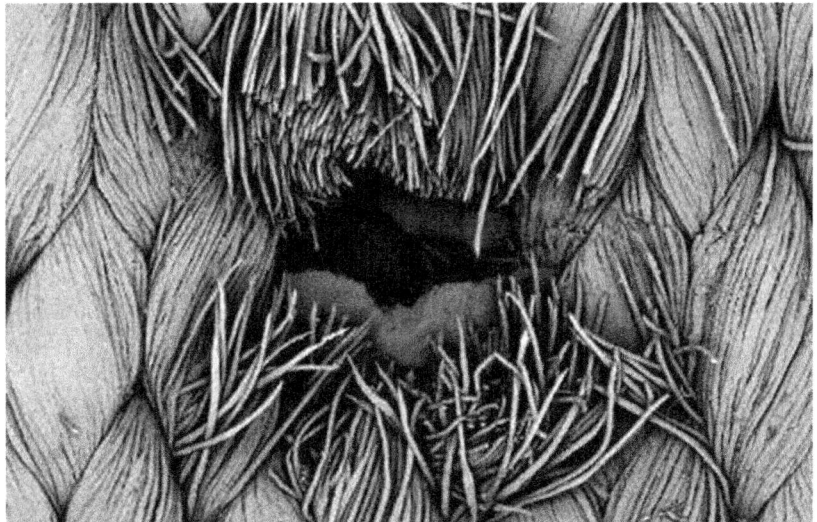

FIGURE 9.10 Pinhole in a multifilament cloth. This could have been caused by a stray sharp particle, a piece of weld splatter or a sharp edge on the filter equipment, e.g. the burr in Figure 9.9.

A large-mesh support cloth (backing cloth) can also be fitted between the filter cloth and the support grid on many filters. This will shield the filtering cloth from sharp or rough edges and may even allow the filtrate to flow slightly more freely.

9.4.2 Cloth Cleaning

While there are several possible cloth cleaning possibilities, belt filters and automatic tower presses have the built-in design advantage of a moving filter cloth that can be passed through a stationary cleaning station and cleaned on both sides relatively easily (Figure 9.11). Other arrangements can be used, although these may be more complicated. For example, Figure 9.12 shows an arrangement in which a moving spray bar travels through a filter press, one plate gap at a time.

In other situations, chemical cleaning agents (e.g. acid or caustic) can be used to remove solid material that is embedded into or precipitated onto filter cloths. The effects of cloth cleaning can have a dramatic effect on the success of a filtration process (Figure 9.13).

9.4.3 Cloth Repair

If a cloth suffers a pinhole one day into an expected lifetime of 100 days, it makes sense to try to patch that hole. If, 95 days into the lifetime, a number of pinholes start to appear, it is time to change the cloth.

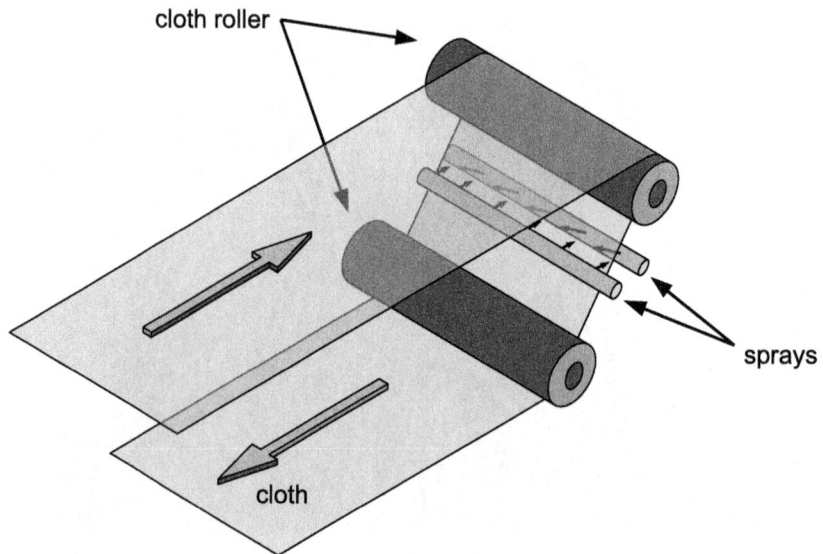

FIGURE 9.11 Cloth wash sprays.

Chapter | 9 Filter Cloth

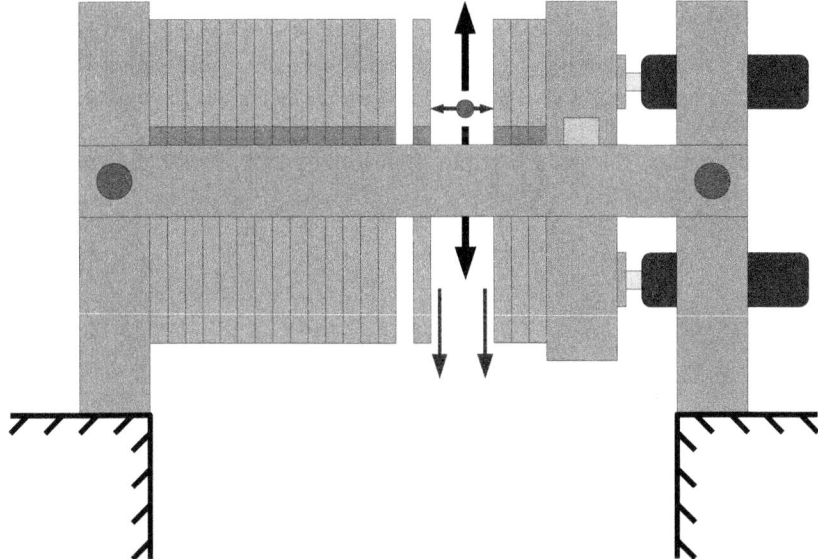

FIGURE 9.12 Filter press cloth cleaning system.

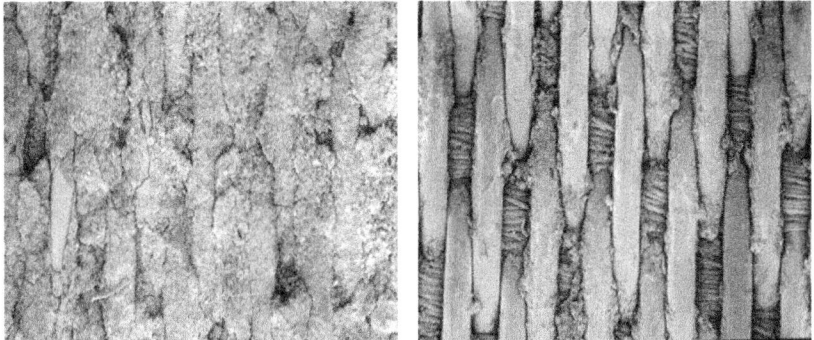

FIGURE 9.13 Before and after cleaning (Clear Edge Ltd).

- Gluing a patch over the hole: This patch may be simply stuck onto the surface of the hole, or (if possible) between the cloth and support grid. The choice of glue is critical, as is the method used. Normally the cloth needs to be dried thoroughly (e.g. with a hot-air gun) before the glue can be used. The glue also needs to dry thoroughly after application, so more production time can be lost.
- Welding a patch over the hole: Some filter cloths can be welded, using either heat or ultrasonics. This method has the advantages that the cloths do not need to be completely dry before the patch can be applied and, once the patch is in place, the filter can be used straight away.

9.5 SUMMARY

The cloth is the single most important component on a solid–liquid filter and usually the most expensive. There is a vast choice available, and cloth selection must balance cost and performance.

Steps can and should be taken to sustain cloth performance and protect the cloth from damage.

Chapter 10

Filter Maintenance

Chapter Outline
10.1 Particular Issues with Filtration Equipment 148
10.2 Speed Versus Machine Sympathy 149
    10.2.1 Optimal Operating Regime for a Multiple-Filter Installation 149
10.3 Component Plant Trials 151
10.4 Summary 151

There are simply too many filter types (and the differences between them are too great) to be able to give a universal maintenance list covering all of them here. The supplier of a particular piece of filtration equipment will have standard maintenance recommendations, but these will probably need to be adjusted according to the precise operation; for example, a particularly harsh or benign process and may imply overmaintenance or undermaintenance. Since filtration plants can be dusty, and since many of the filters use electrical motors and hydraulic/pneumatic actuators, it is important to make sure that these are well cared for and that hydraulic oils are constantly monitored and air quality is maintained.

However, it is worthwhile spending some time reviewing the general issues and points of concern for filtration equipment. Chapter 7 includes a section on some of the particular maintenance issues for each filter type.

Failed components on a filter can lead to harm on three levels:

- Harm to the process: A worn cloth or leaking slurry-feed valve will affect the performance of the process. In general, if these problems are put right then the process performance will resume immediately to its previous level.
- Harm to the equipment: The motive force for dewatering, applied by the filter area, can give a very large pressure. Any small proportional pressure difference between chambers or filter elements can result in permanent bending. Equally, if an abrasive slurry bypasses a worn rubber seal then the resulting jet can burn a wormhole though the plate, even through stainless steel.

- Harm to persons: A spray of slurry coming from a pressure filter or a catastrophic mechanical failure gives the potential for significant harm to people nearby.

There is much to be gained by establishing a maintenance regime in which all consumable parts are scheduled to be replaced before their anticipated lifetime. There will be occasional premature failures (due to supplier quality, process temperature or density excursions), but over a longer period, the possibility of harm will be minimized.

Steps can also be taken to extend the lifetime of these consumable and spare parts.

10.1 PARTICULAR ISSUES WITH FILTRATION EQUIPMENT

On many productions sites, the filters are seen as the biggest maintenance problem. Most filtration equipment would run forever if you were passing clear distilled water through them, but add abrasive, sharp particles that grind into cloths, bearings or valves and that form filter cakes which harden to a concrete-like form and the problems begin. For example:

- Scenario 1: A solid lump of hardened cake dropping into a discharge scraper or discharge screw may instantly hole a filter cloth. Once a small hole has developed in a filter cloth, it can quickly grow and a great deal of solid material will pass into the filtrate. If solids are the product and the filtrate goes to waste treatment then productivity will fall as treatment costs increase. The filtrate system may not have been designed to cope with this amount of solid material, and pumps, valves and pipes could be harmed if this situation is not put right.
- Scenario 2: A piece of cake lodges on the sealing surface of a chamber filter press. This damages the seal and, over the next few days, slurry starts to bypass the seal and burns a wormhole through a stainless steel frame structure. The filter must be taken offline while the steel plate is repaired by welding.
- Scenario 3: For the want of a nail the shoe was lost, for the want of a shoe the horse was lost ....

The combination of abrasive slurries, hardened filter cakes, dust and, often, large pressure differences means that filters have a large inbuilt potential for unreliability. By their very nature they may require more attention than other plant equipment.

A number of steps can be taken to alleviate this situation. If cake or dust tends to accumulate in a particular part of the filter then adding deflector plates or cleaning sprays could make a significant difference. When a filter is installed, it is recommended that a postcommissioning review and, if necessary, small modification projects are used to deal with any issues that arise.

If you are performing maintenance near to a filter, any grinding or welding nearby can produce a spray of hot or molten metal fragments that will fall harmlessly onto virtually every other component except filter cloths, which will be damaged beyond repair.

10.2 SPEED VERSUS MACHINE SYMPATHY

The following section looks in particular at rotary vacuum filters, but some of the points apply to other filter types.

A filter cloth is often at its most vulnerable when it faces a new batch of slurry. So a cloth panel rotating quickly through a trough at 3 rpm will be exposed to an oncoming rush of particles 4320 times per day, whereas at 2 rpm, it will be 2880 times per day. In addition, at the lower speed the filter cake will be thicker and the filtrate will tend to be clearer (and if the particles are abrasive then cloth damage will be reduced). If we include the effect of the extra mechanical stresses of cake discharge at the higher speed (e.g. a cake scraper), it is apparent that the cloth lifetime will be significantly reduced.

Some maintenance engineers work on the assumption that wearing components last as long as the inverse of the speed squared. So, in this example, say the cloth lifetime is a reasonably consistent 90 days at 2 rpm, it would drop to 40 days at 3 rpm. In addition, as components are driven harder, the predictability of their failure becomes more difficult. In the above example, there will be more scatter (i.e. unexpected cloth failures during a weekend) at 3 rpm that at 2 rpm.

In the meantime, the extra capacity gained from increasing the speed from 2 to 3 rpm will be relatively modest, at around 20%.[1] If you drive a car consistently along motorways at 3000 rpm in fifth gear then it will probably give reasonable reliability. The same car driven at 6000 rpm in third gear will suffer more reliability problems and is more likely to break down, with relatively little gain in performance.

This information may be interesting, but not much immediate use if you only have one filter that must be operated at 3 rpm to reach the minimum required capacity. However, if you are running a number of filters it could help to guide the way that you choose to operate and maintain these units.

10.2.1 Optimal Operating Regime for a Multiple-Filter Installation

Say there are ten rotary vacuum drum filters installed on a plant filtering an abrasive and aggressive slurry. The required capacity can be reached by running five of these filters at their maximum achievable speed (say 5 rpm). What would be a sensible operational regime?

---

[1] The capacity of a trough-fed rotary vacuum filter (e.g. vacuum disc or vacuum drum) is, at best, proportional to the square root of the speed (see Section B.1 in Appendix B). So, doubling the speed of one of these units will, at best, give a capacity increase of about 40%, and reducing the speed by half will still give a capacity of around 70%.

Given the relationship between speed and capacity, it is possible to tabulate the scenarios for capacity. Given that the filter cloth is at its most vulnerable when the cake is peeled from it by the discharge roller (or discharged by blow-back air) and the cloth enters the trough to be hit by fresh slurry, we could suppose that the cloth wear is proportional to the number of cake discharges (or, as above, proportional to the speed squared). So the lifetime will be the inverse of this. It is also likely that the wear on some other parts on the filters (e.g. the control head) will increase sharply with speed as they approach their design limits.

Table 10.1 gives the approximate rotational speed required for the number of filters in operation. If seven filters are operated (just two more), the required speed is around half of that for the original scenario of five filters. In addition, the total number of filter rotations (which we suppose determines the spares consumption) reduces sharply, with a reduction of nearly 30% with seven filters versus five.

This is not to say that the optimum will always be to run all ten filters at 1.3 rpm. Other issues need to be considered:

- Will the increased cake thickness spoil the cake discharge?
- Will cake cracking become an issue at this slower speed?
- How would any washing processes be affected?
- What will happen to the overall vacuum pump capacity and power consumption?

In a scenario like the one above, in which you are trying to establish an optimum operating regime in a multiple-filter installation, a suggestion would be to run one of the filters at a slower rate and monitor closely its performance (throughput, washing result and cake discharge), spare-part lifetimes (particularly cloth) and power consumption. The results from a trial like this can then be used to make informed decisions about how to maximize the lifetime of spares.

TABLE 10.1 Operational Scenarios

No. of filters	Required speed (rpm)	Total no. of rotations per minute (no. of filters multiplied by rpm)
5	5	25
6	3.5	20.8
7	2.6	17.9
8	2	15.6
9	1.5	13.9
10	1.3	12.5

In short, with speed the cost of spares and the likelihood of unexpected shutdowns increase more rapidly than the increase in capacity.[2]

As always, there can be no hard-and-fast rules, because of the huge variety of filtration applications. However, any decisions should be better informed by an understanding of the relationship between speed, capacity and running costs for a particular process. This information may even provide the justification needed to buy another filter for the plant.

10.3 COMPONENT PLANT TRIALS

Plant trials are usually embarked upon with a great deal of enthusiasm, with the hope that something will improve but, in time, this enthusiasm can be lost and replaced with an occasional embarrassed shrug before being forgotten about. Some extra steps can be taken to avoid this outcome.

- Components get covered in slurry and dust: Make them highly visible; add a clear tag to each new component.
- Make daily inspections/measurements: Make the results of these public.
- Inform and inspire the team: Show how the change fits into the process (the success factor tower can be a useful tool). Prequalify the criteria for success, making use of the phrase "so that …":
  - "We will increase the throughput by 10% … *so that* we can increase our company sales to meet demand from a new market".

10.4 SUMMARY

Filtration machinery faces particular maintenance challenges because of the abrasive nature of many slurries and the difficulties in handling filter cakes.

Dust and filter cakes should either be prevented from reaching vulnerable parts of the filters (bearings, seals and cloths) or cleaned away regularly. Either a regular inspection and cleaning procedure or, if necessary, small modifications to the filter (deflector plates or spray bars) should be used for this.

In multiple-filter installations, the compromise between running fewer filters at a faster speed or more filters at a slower speed should be investigated. It may be useful to consider the number of cycles rather than the number of hours in operation (or at least to take into account the speed of the filter).

---

[2] One client I worked with in the food industry had offered bonuses to the team who could produce the most product in one shift. Production records were indeed broken, as were many of the critical components on their filters. The result was many shifts of lost production.

Chapter 11

Filter Operation

Chapter Outline
11.1 Operational Choices 154
    11.1.1 Example: Trough-fed Filter Capacity Versus Speed 154
        11.1.1.1 Multiple Filter Scenarios 155
11.1.2 Effect of Speed on Top-fed Filters 157
11.1.3 Batch Pressure Filters 157
11.2 Summary 158

As with all of the success factors listed so far, the way that you operate a filter can affect and possibly spoil the success of a process, even if everything else is in order.

That said, out of all of the success factors it may be the simplest to change and sometimes be done immediately, with the appropriate agreement of everyone involved. The danger is that this can lead to a culture in which every operator or process engineer has his or her own opinions over the operation of the filter, resulting in constant changes being made, sometimes between shifts. It can become very difficult to keep control of a filter plant if the cycle times or rotational speeds change hourly.

If changes are made then they should come one at a time so that the results of these changes can be properly attributed. You may make three changes; one may have a highly positive impact on filtration outcome but this may be masked by a negative impact from the other two changes. Then, all three are never tried again.

The operational parameters include:

- pressure or vacuum level
- speed
  - rotational speed of trough-fed vacuum filters
  - linear belt speeds
  - centrifuge spin speed
- filtration stage times (feeding, washing, air drying, etc.)
- wash liquid addition points (if movable) and wash liquid volumes
- slurry level in drum filters.

The effect of many of these parameters can be tested at a small scale and finding the optimum may involve trial and error (informed by ideas about *what could be happening* within the filter cake and slurry at a microscopic scale).

Equally importantly, it is useful to know how to deal with excursions from the normal process conditions: How do you process a contaminated batch of fermentation broth? How do you handle a batch of a badly crystallized intermediate product that is contaminated with very fine material?

One of the most useful approaches is to appoint a small group of people (or possibly one person) who are well informed about the basic ideas of filtration theory, but who also understand how the filtration performance affects the performance of the overall process. They should be able to notice any signs of deviation from the normal process conditions (the nature of the cake, cracking, wash water distribution, cloth condition, etc.).

Over the past decade, more attention has been paid to control systems that respond to filtration conditions. Online measurement of cake thickness, wash filtrate concentration or even cake color can be used to modify, for example, the speed of a vacuum drum filter or the stage times of a pressure filter cycle.

However, while these advances in instrumentation are useful, they should not replace the wise head with responsibility for the operation of a filter.

11.1 OPERATIONAL CHOICES

In Chapter 3 (with mathematical background in Section B.1, Appendix B), it was found, under conditions of constant pressure and with all else being equal, that the volume of filtrate collected through a growing filter cake is approximately proportional to the square root of the time:

$$V \approx k \times \sqrt{\text{Time}} \qquad (11.1)$$

where k is a constant of proportionality that includes differential pressure, filtration area, liquid viscosity, cake permeability and, possibly, factors to take cake compressibility into account.

11.1.1 Example: Trough-fed Filter Capacity Versus Speed

For trough-fed filters (vacuum drum, vacuum disc, etc.), the capacity, C, will be close to, but probably slightly less than proportional to the square root of the speed:[1]

$$C \lesssim \sqrt{\text{Speed}}$$

---

[1] The reason that this may be slightly less is to do with machine characteristics. Imagine that there is no filter, just a control head rotating in a large-diameter pipe. If the control valve is stationary then the flow through the pipes will become fully developed. At very high control valve speeds, the flow rate (and therefore the capacity of a filter) will be reduced. In addition, it will take a certain amount of time for vacuum to evacuate the air from within each sector and associated piping, time that is not beneficial in filtration terms.

So, at higher speed, each sector will discharge less cake, but more often.

11.1.1.1 Multiple Filter Scenarios

Suppose that you have three vacuum drum filters installed to dewater a typical, reasonably coarse, mineral product.[2] You normally operate two units, with one on standby. The two units deliver the required capacity at 2 rpm, and the cake thickness is about 10 mm. Would it be better to run three units at the same time, taking one off at a time for routine maintenance?

First of all, it is impossible to answer this question definitively without further information, but let's look at some of the implications of this change that would feed into a decision-making process.

The capacity, C, of each unit is approximately proportional to the square root of the speed. So, overall:

$$C \propto \text{Number of units} \times \sqrt{\text{Speed}}$$

So, in this case:

$$C \propto 2 \times \sqrt{2}$$

and, in the proposed situation:

$$C \propto 3 \times \sqrt{\text{New speed}}$$

So, since we are looking to have the same capacity:

$$3 \times \sqrt{\text{New speed}} = 2 \times \sqrt{2}$$

$$\text{New speed} = \left(\left(\frac{2}{3}\right)\sqrt{2}\right)^2 \approx 0.9 \text{ rpm}$$

This is a much bigger reduction than you might have expected if you are new to filtration.

Also, since:

$$\text{Cake thickness} \propto \frac{1}{\sqrt{\text{Speed}}}$$

in this example:[3]

$$\text{New cake thickness} = \frac{10 \times \sqrt{2}}{\sqrt{0.9}} = 15 \text{ mm}$$

---

[2] Assumed to be incompressible.
[3] In fact, because cake discharge is continuous, and assuming that the cake is incompressible, this reduces to:
$$\text{New cake thickness} = \text{Old cake thickness} \times \frac{\text{New number of filters}}{\text{Old number of filters}}$$

TABLE 11.1 Vacuum Drum Filter Scenarios

No. of units	Speed (rpm)	Cake thickness (mm)
1	8	5
2	2	10
3	0.9	15
4	0.5	20

The results of these calculations are given in Table 11.1 (with the implications of running one and four filters included for good measure).

A thicker cake normally produces a more clear filtrate; so, if clarity is a problem for you, then running three filters could improve the situation.

Chapter 10 reviews the implications of filter speed on spare part lifetimes, but if we assume that the lifetime of the cloth is related to the number of times that each panel on the filter discharges cake per minute, then there is a good argument that the cloth lifetime will increase in this new scenario, since before there were $2 \times 2 = 4$ discharges per minute and now there are $3 \times 0.9 = 2.7$. So, if other spare parts follow this rule (and it is possible to argue that we are underestimating the improvement in spares lifetime) then the overall consumption of spares would actually drop if more filters were used!

There are now three filters connected to the vacuum system, so we should expect the air consumption to increase. However, this is also worth checking. First, there will now be less air consumed to evacuate the panels and piping at the beginning of the rotation. Also, the air flow through the cake will reduce per unit area, because the cake is thicker. In fact, if we assume that the air flow is inversely proportional to the cake thickness then the air flow ought to be roughly the same, but this would need to be checked because other factors may come into play, for example cake cracking.

If this is a washing application, the potential washing time will more than double with the third unit running, so it may be possible to improve the washing result. If you were previously struggling to make the required product quality, then you may find that it is now easier. This is borne out by, for example, work on red-mud filters by Borges and Aldi (2009), who reported that "a rotation of 1.4 rpm resulted in an average caustic concentration of 17.7 g/L, against 23.6 g/L when operating with 3.1 rpm".

It is more difficult to estimate the effect of all of this on the final cake moisture. There will be a lower flow of air through a thicker cake, for longer. It can be argued that the outcome for cake moisture would be broadly neutral. However, this does not take into account unknowns such as cake cracking.

Laboratory tests will provide further validation of the above estimates. It may also be possible to check the air flow through the cake at this scale (although there may be issues of scaling up).

Taking all of this into account, and after checking at laboratory scale, there are probably enough benefits in terms of production cost and, possibly, production quality (if filtrate clarity or wash results are critical) to warrant a plant trial running three filters instead of two.

The biggest opposition to this idea might come from your maintenance department who, at first glance, will just see more work, from more filter-hours of operation. It would be a good idea to turn the argument around to the total number of filter revolutions per minute: 2.7 versus 4.

11.1.2 Effect of Speed on Top-fed Filters

The capacity of pan filters and horizontal belt filters is determined entirely by how much slurry is poured onto them. What is less certain is whether there will be time for the filtration and drying as well as any washing process to complete. The danger is that the filter discharges only partially dewatered slurry.

However, for a filter that is capable of completing all of the steps (i.e. has enough area), there remain choices over the operational speed. While it will make no difference to the capacity, it is possible that the performance of the process could be affected, for better or possibly worse.

First of all, there are other reasons why reducing the speed of a pan or horizontal belt filter could benefit the process. One is the well-rehearsed argument about spares lifetime from previous sections.

Reducing the speed can potentially affect the uniformity of the filter cake. There may be disruption to the surface of the cake, for example from wash water. Say that this disruption causes 5mm ripples across the surface of the cake. These would give a significant variation in properties for a cake (and potentially washing and air-drying results) with an average thickness of 15mm, but less so for a cake that is 30mm thick.

Finally, reducing the speed will reduce the load on the discharge system somewhat. The same amount of cake, at approximately the same cake moisture, will be discharged per second, but the speed of cloths against scrapers (or discharge screws) will be reduced and the kinetic energy (that will be dissipated through wear) will be reduced significantly. Furthermore, the specific design of scraper could perform optimally at a certain speed and it is therefore possible that the scraper type could be changed.

11.1.3 Batch Pressure Filters

There are operational choices to be made for batch pressure filters. For filtration processes with highly compressible cakes, there may be a Goldilocks pressure (not too low, not too high) at which the capacity is highest; testing will tell you where to look.

Variable volume chamber filters (i.e. membrane filter presses, tube presses) may allow you to measure the cake thickness from cycle to cycle by measuring the amount of pressing fluid needed to press the cakes (this is especially convenient if water is used in a closed loop).[4] Cake washing results can be highly sensitive to cake thickness, especially if the wash liquid is pressed through the cake with a diaphragm. The optimum thickness can be determined and the slurry pumping time set accordingly.

If you have a number of batch pressure filters then it makes sense to share feed pumps, compressors and other utilities. In this case, the units should be sequenced so that they are not all demanding (say) slurry feed, or compressed air, at the same time.

11.2 SUMMARY

Filtration outcomes can be significantly affected by the way in which a filter is operated.

With the full agreement of the people involved, it is possible to make changes relatively quickly. These changes, and their effect on filtration outcome, should be carefully tested and monitored.

---

[4] It may also give an early indication of diaphragm or bladder failure if water is lost from the otherwise closed circuit.

Part IV

Creating and Sustaining a Competitive Advantage

Chapter 12	Process Testing	161
Chapter 13	Getting the Most from Filtration Processes	169
Bibliography		181
Appendix A	Useful Expressions	183
Appendix B	Flow Through a Growing Porous Filter Cake	189
Appendix C	Forms and Templates	195
Appendix D	Sample Test Method	199

Chapter 12

Process Testing

Chapter Outline

12.1 Test Equipment	162	12.6 Data Acquisition	165
12.1.1 Laboratory Scale	162	12.7 Archiving Data	166
12.1.2 Pilot-Scale	163	12.8 Cake Washing	166
12.2 Testing Program	164	12.9 Analysis	166
12.3 Design of Experiments	164	12.9.1 Scaling Up	168
12.4 Sampling	165	12.10 Summary	168
12.5 Example Method	165		

> ... at the heart of science is an essential balance between two seemingly contradictory attitudes – an openness to new ideas, no matter how bizarre or counter-intuitive, and the most ruthless skeptical scrutiny of all ideas, old and new ... science requires the most vigorous and uncompromising skepticism, because the vast majority of ideas are simply wrong, and the only way to winnow the wheat from the chaff is by critical experimental analysis.
>
> Carl Sagan, *The Demon-Haunted World*, Ballantine Books, 1996.

Physical testing, in a laboratory or at pilot scale, is an essential part of understanding a filtration process. In the future, it may become possible to input information on a solid–liquid process (particle size, shape, cloth geometry, zeta potential, etc.) into a simulation model that will deliver throughputs, moistures and washing results. For the next few years, at least, however, the only practical way to obtain this information will be through testing, subjecting the slurry and cloth to similar conditions, albeit at a smaller scale.

Within the range of testing, two main types emerge: first, to determine the filtration parameters (e.g. the cake resistance) so that a simulation of various full-scale operating scenarios (or even filter types) can be assessed; and second, to mimic the operation of a particular full-scale filter. Further discussion of filter parameter acquisition and simulation is given by, for example, Wakeman and Tarleton (1999), Stickland et al. (2011) and Nicolaou (2003).

Testing of slurries during process development and equipment selection/ plant design is vitally important, and the sooner these tests are done the better.

These tests should give the filtration process outcomes to a good level of certainty, including cake moisture, washing results and operating costs. Tests can also be used to deliver qualitative information, such as the tendency for cake cracking or information on how the cake discharges from the filter cloth. Test reports and data may also form a part of any process performance guarantee from an equipment manufacturer.

Testing provides an opportunity to try out new ideas for an existing filtration process. In process optimization, it is better to be wrong with a few liters of product than with an entire day's production.

In general, it is more convenient to produce data at a laboratory using vacuum or pressure filtration than a centrifuge. However, it is important that the test engineer is able to incorporate the features of large-scale equipment into the testing procedure, for example a pause between filtration and cake washing.

This chapter looks at the essential requirements of filtration testing.

12.1 TEST EQUIPMENT

While there is a large variety of test filtration equipment available, it broadly fits into two categories: laboratory and pilot scale.

12.1.1 Laboratory Scale

These small, portable units (e.g. Figure 12.1) can be used to mimic the conditions in full-sized units, albeit sometimes with compromises of scale. They are usually designed to be used on a bench in a reasonably equipped laboratory and can provide essential information throughout the lifetime of a filtration plant, even during its design. There may be some inherent compromises but, overall, given the quantities of slurry needed for testing and the amount of information that can be produced, this is the best place to start.

Every plant with a large-scale filtration process would benefit from having a laboratory unit that can be used at short notice to diagnose plant filtration issues and to assess ideas for process optimization. It is better to find out that a new cloth material is not suitable after challenging it with a liter of slurry in the laboratory than to spend hundreds of euros buying and installing the new cloth on the plant, only to find out that the filtrate is too cloudy and a day's production has been lost.

Ideally, the laboratory equipment should be simple to use and should accept the same filter cloth as the full-scale unit that it is mimicking.

The vendors of filtration equipment generally offer units like these, although they may be designed specifically to mimic their own large-scale machines (a company that supplies vacuum filters is unlikely to offer a laboratory pressure filter and vice versa).

Typically, a quick laboratory trial can be planned and performed in a matter of hours, whereas a pilot-scale campaign can involve several weeks of planning and execution.

Chapter | 12 Process Testing   163

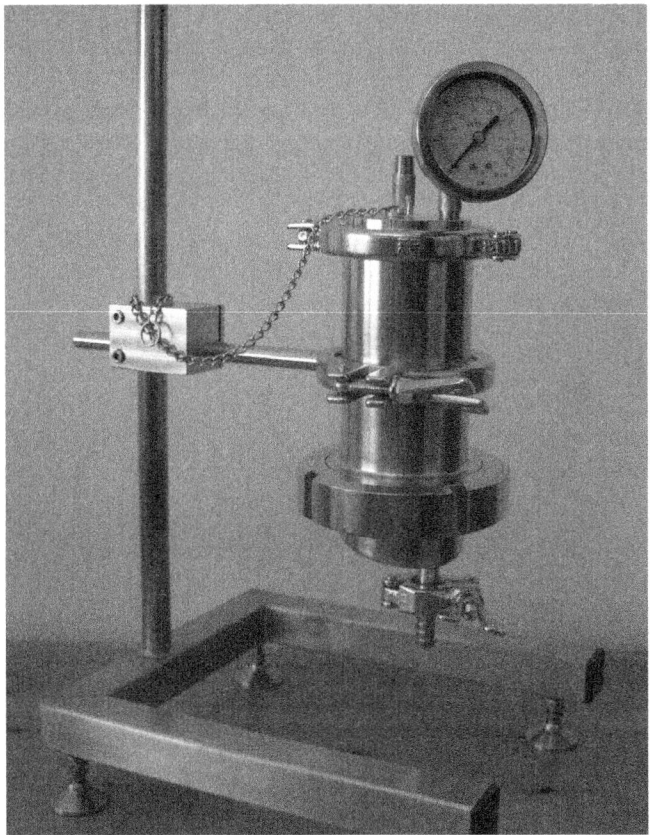

FIGURE 12.1 Laboratory test filter: this unit can operate as a pressure filter or a vacuum filter.

12.1.2 Pilot-Scale

Using pilot-scale equipment provides the opportunity to perform extended trials, which may expose longer term problems, for example cloth blinding, and also to find out about equipment-specific issues (e.g. how a particular slurry spreads over a vacuum belt filter and how uniform the cake thickness will be).

Typically, pilot-scale trials are used during plant design and most large-scale filtration equipment suppliers have pilot-scale units for hire, as it helps them to validate their claims. These units normally use exactly the same operating principle as the full-scale plant and can come skid-mounted complete with slurry tanks, pumps, piping, control system and compressors.

If the pilot-scale trial is used to assess possible equipment for the full-scale plant then it is a good opportunity to involve the people who would use the production plant, to introduce them to the concepts and listen to their feedback. This opportunity is not always taken.

These tests will be quite involved, can easily cost tens of thousands of euros and take several weeks (in effect, they involve building a full filtration plant). It is worth taking the time make the most of them.

In most circumstances, it would not be practical to buy these units, unless a large number of processes are using similar equipment. However, some universities or other institutes may have pilot-scale equipment that can be used for a long-term trial.

Pilot-scale test units may also be used to generate significant quantities of filtrate or filter cake to be used in, for example, pilot-scale precipitation or cake drying trials.

12.2 TESTING PROGRAM

The testing procedure should be documented, unambiguously and thoroughly, so that any other person can reproduce the exact testing conditions later on.

The contents of this document should include:

- Objectives: The overall purpose for the test, where possible to be framed in terms of the overall success of the process (e.g. "To evaluate the effect of chemical filter aids on final cake moistures. Each percentage point improvement in moisture reduces fuel consumption by …").
- Apparatus: A clear, numbered, list of equipment and tools (hardware and software) that were used.
- Set-up: Ideally, a labeled photograph of the experimental set-up, including detailed photographs where needed (e.g. a close-up of a valve setting or pressure gauge).
- Sampling: Procedures for sample collection, transportation, storage and preparation, including handling procedures, temperatures, etc.
- Method: An entirely unambiguous description of the steps needed, ideally as a numbered list.
- Data sheet: A standard form, either on paper or as a spreadsheet.
- Analysis: How the data that have been collected can be used to assess the performance of the filtration process.
- Conclusions and recommendations.

12.3 DESIGN OF EXPERIMENTS

Even for a relatively simple filtration process, such as dewatering in a batch pressure filter, there will be a large number of possible combinations for testing. You may choose to investigate three of the following:

- feed times
- pressures
- pressing times

- air drying pressures
- air drying times.

This will result in a very large number of tests, if you vary one at a time and keep the others constant. There are software tools available for the design of experiments and multivariate data analysis that will crunch down the number of experiments needed and therefore the time and overall slurry sample volume needed; see for example Häkkinen et al. (2008).

The number of experiments can be reduced by using knowledge of the way that the cake forms (and remembering the rate of cake formation under constant pressure, for example). Tests can also be simplified using rules; for example, stop pressing once the pore pressure has drained from the cake (in other words, when there is little benefit to be gained from continuing) or when the wash filtrate concentration reaches a certain level.

12.4 SAMPLING

The nature of a sample can change significantly during storage. Precipitation, coagulation or microbial attack (bacterial or fungal) can all have an effect.[1]

In general, and where possible, the sample should be tested as soon as possible and should be kept in suspension. In any case, the sampling, storage and transportation procedures should be carefully thought out and noted in the report.

12.5 EXAMPLE METHOD

Where possible, the method used should mimic the full-scale operation as closely as possible. For example, if you pause before adding a wash liquid so that the cake dries and cracks, then the washing information generated will be of limited value.

It is also essential to display sympathy towards the structure of the filter cake. It is better if the cake remains under pressure or vacuum at all times so that it does not shift and its structure remains intact. If this is not possible, then the compromises should be minimized and, where they exist, reported.

12.6 DATA ACQUISITION

The way in which the test data are acquired may depend upon the available time and equipment. If the process steps (e.g. cake formation, cake washing and air drying) all take several minutes then it should be possible simply to write down data directly in a data sheet (e.g. filtrate weight or volume, airflow) as you go.

---

[1] I once sent 1000 liters of slurry to Finland for an extensive trial without explicitly specifying that it should be transported in a heated lorry; the resulting cubic meter of ice and calcium carbonate took several days to thaw.

However, if the process is more rapid then some other form of data acquisition will be needed.

Most weighing scales used to measure filtrate weight have an optional system that can be used to collect data onto a computer.

Another possibility is to use a camera to video the key data as they are being acquired. By arranging the timer, filtrate weight scale display and pressure gauges so that they are in the same shot, you can video the whole test and review the video later, pausing it every so often to write down the numbers from the screen. This offers a quick and relatively easy way to acquire good quality data and I have been able to use this method on a filtration process lasting for about ten seconds. Using a card with the date and test number like a clapperboard at the beginning of the test will create a permanent record. Since most cameras also have a microphone, comments or descriptions (e.g. the time at which wash water was added, or noting that the cake cracks) can be added along the way.

12.7 ARCHIVING DATA

Recording of data is a perfect application for spreadsheets and Figure 12.2 shows an example.[2] The quality of the data produced can be checked using a mass balance across the filter; a significant scatter in these mass balances could betray a problem in the methodology or execution of the tests. File names should be clear and the data stored and backed up.[3]

12.8 CAKE WASHING

Cake washing can be tricky to accomplish at the laboratory scale. However, as long as the test is arranged well in advance, and there are enough prelabeled containers for the various wash filtrates, it is more than possible to investigate complicated multistage counter-current washing regimes.

12.9 ANALYSIS

The report should contain all of the process pressures, vacuum level and filtration cycle stage times or simulated belt speeds/rotational speeds. It is a simple matter to create cell formulae in the data sheet to calculate useful parameters. To give an example, one particularly important outcome is the filtration capacity per unit area, per hour, for a batch filter, which in terms of dry solids is:

$$Capacity = \frac{mass\ of\ solids\ in\ cake}{test\ filter\ area} \times \frac{60}{total\ cycle\ time\ (mins)} \qquad (12.1)$$

---

[2] This can be downloaded from http://www.solid-liquid-filtration.com
[3] It can be useful to date stamp the data files by appending the date of testing in the form YYYY MM DD, so that the files will be in date order when organized by name on the computer.

Chapter | 12   Process Testing

			Test number							
			1	2	3	4	5	6	7	8	9

Test Data — Date: — Performed by:

Project
Objective
Equipment
Area

Cloth

Slurry
- % solids wt/wt
- temperature °C
- pH
- Density gpl
- Comment

Filtration
- time s
- pressure bar
- volume ml
- filtrate solids ppm

Washing
- liquid name -
- time s
- pressure bar
- volume ml
- filtrate solids ppm

Pressing
- time s
- pressure bar
- volume ml
- filtrate solids ppm

Air drying
- time s
- pressure bar
- Flow-rate lpm
- volume ml
- filtrate solids ppm

Cake
- weight Kg
- thickness mm
- moisture %w/w

Filtrate
- Volume ml
- filtrate solids ppm

Comments

- Total time s
- Cake solids Kg
- Capacity kg/(m²h)

General comments

FIGURE 12.2 An example of a test data sheet. The spreadsheet can be used to give a mass balance across the process for solids and liquids to give an indication of the quality of the data.

The total cycle time should include the time for each stage of the process, as well as a time to allow for the technical operations of the filter (cake discharge, valve operations, pressure release, etc.):

$$\text{Total cycle time} = t_{\text{filtration}} + t_{\text{washing}} + t_{\text{pressing}} + t_{\text{drying}} + t_{\text{technical}} \quad (12.2)$$

There are equivalent calculations that can be used to give the capacity of continuous filters and these are included in the downloadable spreadsheet.

12.9.1 Scaling Up

In general, it should be safe to scale up by area. However, two main issues should be considered.

There may be physical compromises with the test equipment; for example, if a small diameter test filter is used, there is a chance that edge effects may compromise the results. In general, the diameter (or smallest length for a rectangular unit) of the filtering area should be significantly greater than the thickness of the cake produced. This gives a greater chance of noticing any cake shrinkage or cake cracking phenomena.

In addition, it is important to allow for the operating characteristics of the full-scale filter. For a very fast rotating drum filter, an allowance should be made for the time to evacuate each filtrate pipeline connection to the filtering sector, and for liquid flow to develop fully within the pipelines.

12.10 SUMMARY

Small-scale tests, at either laboratory or pilot scale, will provide a great deal of information on the filterability parameters of a particular slurry (so that filtration processes using all manner of filtration equipment can be simulated) or the performance of a particular filtration device with a given slurry.

A great deal of care should be taken in sample preparation, performing, data archiving and reporting of these tests.

Chapter 13

Getting the Most from Filtration Processes

Chapter Outline
- **13.1 Product Development and Process Design** 170
- **13.2 Equipment Selection** 170
  - 13.2.1 Principles of Equipment Selection 171
  - 13.2.2 The Process of Buying (and Selling) 172
    - 13.2.2.1 Relationship Management 173
    - 13.2.2.2 Cloth Suppliers 173
    - 13.2.2.3 References 173
  - 13.2.3 Second-hand Equipment 174
  - 13.2.4 Slurry and Application Questionnaire 175
  - 13.2.5 After Commissioning 175
- **13.3 Process Optimization** 175
  - 13.3.1 Optimization Projects 177
    - 13.3.1.1 Assembling a Team: Appointing a Champion 178
  - 13.3.2 The Tower of Filtration Process Success Factors 178
- **13.4 Summary** 179

Commercial production processes generally have three life-stages:

1. Product development and process design
2. Project execution and equipment selection
3. Production and process optimization.

The majority of process designs are never implemented, usually for commercial reasons. So, while the sales of new filtration equipment of the type described in this book are around €2 billion per year, there may be more than ten times this value of filtration equipment under consideration at any time in various possible process designs. Equally, given that a typical lifetime for a piece of filtration equipment can be around ten to twenty years, there are several billions of euros of equipment in operation.

It makes sense to get the most from this equipment by optimizing the choices made during the three phases of its existence.

Solid-Liquid Filtration. DOI: 10.1016/B978-0-08-097114-8.00013-7
Copyright © 2012 Elsevier Ltd. All rights reserved.

169

13.1 PRODUCT DEVELOPMENT AND PROCESS DESIGN

It is not uncommon for filtration to be neglected during product development and process design: if a small sample filters within a few minutes on a basic laboratory Büchner filter through paper, or if the material to be filtered resembles another that is already in production, then it is assumed that there will be no problem later.

If these assumptions are wide of the mark, this can lead to significant problems during project execution and onwards, into production.

So, the first principle should be to consider the filtration steps in a process as early as possible, even if only to know what it is that you don't yet know:

- Perform whatever testing you can, as soon as possible. Even if it is with all of the slurry that you have (e.g. from a beaker-scale crystallization or laboratory flotation of a core sample), it will be ideal to learn something about:
  - capacity per unit area for vacuum and pressure filtration and even possible candidate filter designs
  - suitable (and unsuitable) filter cloths, in terms of filtrate clarity and cake discharge
  - final cake moistures, washing results and wash liquid consumption.
- Make initial contact with filtration equipment and filter media suppliers as soon as possible. Even if it is a long time before you will be considered a sales prospect, there may be masses of information to be learned from their customers' similar processes.

While most product development projects do not go all the way to project execution and, ultimately, production, if you have a ready-made knowledge bank for the filtration step then you will reduce the uncertainty (and likelihood of mistakes), as well as the time and effort required when a project does ramp up.

Even if the best that you can do at this early stage is to establish the range of likely performance, this information can be used in a sensitivity analysis of the viability of a process.

13.2 EQUIPMENT SELECTION

When you choose a piece of filtration equipment, you are making a commitment, on behalf of your company and the people who will operate, maintain and rely upon it for many years to come.[1]

As discussed in Chapter 5, this decision will affect the competitiveness of your process, in the familiar terms of production cost, product quality, safety, health and the environment (SHE) and productivity. Let's say that your process

---

[1] In 2002, I saw a filter press that had been in continuous operation since 1961, even though it was second hand then and had, in fact, been manufactured in the 1940s. It has since been decommissioned.

is producing 10 tonnes of product per hour for 8500 hours per year, then a change (positive or negative) of €1 per tonne (say it reduces or increases your drying costs) will come to more than €15 million over a twenty-year lifetime for the process.

Clearly, then, it is important to get this decision right.

For some reason, the choice of filtration equipment can be an issue surrounded by particularly human responses. I have met more people who are either delighted or miserable about their filtration process than people who are just reasonably happy. There are a number of possible reasons for this:

- The filtration process is often the point at which a solid product emerges, blinking, into the daylight for the first time in a process, having been precipitated, settled, pumped and classified in closed pipes or under the surface in thickeners. So it is often the point at which a problem in (say) precipitation becomes known, perhaps as a very wet and sticky filter cake that will not convey.
- Filtration equipment, especially if it adds a great deal of value to a process, can be considerably more expensive, per kilogram of steel, than other processing equipment.
- Decisions are often hurried.

This section will look at some of the principles that should be behind equipment selection, before looking at the selection and buying processes.

13.2.1 Principles of Equipment Selection

It is essential that equipment selection is:

- based upon evidence:
  - testing
  - references
  - experience
- considered in terms of the four aspects of processing success:
  - cost, quality, SHE, productivity
  - financially modeled, over the anticipated lifetime of the process
- carried out in an unhurried and timely manner.

The gestation period for a production process is certainly more than a year, and could be as much as five years. It is not unusual for the development to focus on the chemistry and reaction engineering, while filtration is neglected at this stage.[2]

---

[2] Software is available to assist in this equipment selection process; see http://www.filtrationsolutions.co.uk

Instead of arriving at detailed plant engineering with almost no idea of how the filtration will look, it should be possible to have a pretty good idea – which can be reached via:

- in-house studies: review of available technologies and, ideally, laboratory-scale testing.
- third-party study, e.g. a local university or institute
- reference visits
- preproject contact with vendors
- research from other parties: cloth suppliers, other supplies (e.g. chemical additives)

It would be far better to install the best equipment (and cloth) for the duty rather than equipment from the supplier with the best sales process and people.[3] Over the past few years, there has been a trend in the filtration industry towards fewer suppliers, but each with more filter types available than before. With this comes a broader range of filtration options, but perhaps some loss of specific equipment knowledge.[4]

At all stages of the equipment selection process, the tools are available to assess continuously the likely outcomes of the filtration processes (perhaps based upon testing or vendors' claims) in terms that relate directly to the success of the overall process (Chapter 5). The success factors that, together, deliver these filtration outcomes are also known and discussed throughout Part III of this book.

So, take advantage of the knowledge and experience available from filter suppliers, but make sure that you keep as much control of the process as you can.

13.2.2 The Process of Buying (and Selling)

There are a number of layers to selecting the piece of equipment that will eventually sit in your plant (not necessarily in order):

- suitable filter cloth
- suitable driving force for filtration
- suitable filter construction(s)
- suitable vendor(s).

It is helpful to compartmentalize these in your thinking so that you can focus on the important issues within each layer.

Throughout this process, the internet and expert system software can be useful in arriving at a shortlist for further investigation. In particular, the Filtration and Separation Buyer's Guide (at http://www.filtsepbuyersguide.com/)

---

[3] If these things combine, then all is well.
[4] One previous employer of mine went from having a single filtration technology to six distinct technologies (through acquisition) in little over five years.

will help to find vendors of different types of filtration equipment, and the excellent Engineering Aspects in Solid–Liquid Separation page (at http://www.solidliquid-separation.com/) gives detailed information on many filter types. However, these may not pick up on some of the subtleties of a particular situation – there is sometimes no substitute for expert opinion.

13.2.2.1 Relationship Management

Each of the pieces of equipment outlined in Chapter 7 is a large capital purchase and, for most of the available suppliers of this equipment, will represent a major sales target. On an individual level, a filtration equipment salesperson may have fewer than ten active sales cases at any one time, and may have budgeted to sell fewer than this in any particular year. Your order (or lost order) is therefore even more significant to him or her: it could be the difference between a good and a bad year.

Given that the period from first contact to commissioning (after which you may hear much less from the filter salesperson) will be at least a year and possibly more than two years, it makes sense to accept that you will be spending plenty of time in each other's company.

13.2.2.2 Cloth Suppliers

The filter cloth is the single most important component in any filtration process (the filter machinery is simply a mechanism to introduce the slurry to the cloth/grid, to allow the filtrate to flow and to remove the filter cake). While many cloth vendors offer similar products, some may have specialized cloths for specific duties.

There can be no harm in contacting filter cloth suppliers to ask for information on your application, or similar applications. In general, they are in a position to give you their experience without prejudice.

13.2.2.3 References

The typical British house buyer spends a good deal less than an hour looking at their future home before agreeing to pay about five times their annual salary for it (spread out over the next few decades). They rarely try to do the journey between their possible new home and their place of work at rush hour and they rarely introduce themselves to their possible future neighbors before signing the deal. It is a trend that is often repeated in filtration equipment buying.

Request an up-to-date reference list from potential candidate suppliers. Ideally, you should ask for this list to include, at least, the date of supply, customer name and application (if they are not confidential).

Making contact with, and visiting, some of these references is the best opportunity that you will have to learn about the positive and negative aspects of the filter type that you are considering (and of the company that supplied it). The reference visit is one of the red-letter days in a buying process and you should do everything that you can to ensure that the right people are there and that you ask all of the questions.

The worst that can happen is your host asking you to leave early and never come back if you ask the wrong questions (which is extremely unlikely). They are far more likely to give you their honest opinion. As I mentioned in the introduction to this chapter, you are likely to encounter people who are either thrilled or miserable about the filtration process. So, it would be wise to consider all views on a topic and not to dismiss a candidate filtration technology on the word of one particular disgruntled maintenance engineer. It is worthwhile spending time to prepare a checklist for any reference visit (see Appendix C for an example) and making a thorough record of the people seen and their comments.

In addition, there is always a first item on any list, and there can be a significant competitive reward from being early on a supplier's reference list, in that you would benefit from new technology before your competitors. Do not necessarily be put off by the comparative lack of references from one particular vendor.

13.2.3 Second-hand Equipment

Buying second-hand equipment can save a great deal of money. However, it is important to recognize the risks, so that they can be minimized. At best, you will save money; at worst, you will be selling the filter for scrap value, having spent many hours trying to resuscitate it.

The key issue is knowledge of the machinery, and especially who has that knowledge. If you have an existing filter, are looking to expand and find a nearly new, almost identical filter to yours then, clearly, you are in a good position. Your staff will be familiar with the maintenance and operation and it should be relatively simple to reproduce the existing installation. If you are looking at, for you, a new technology, then it is important to gain access to the knowledge and experience.

Second-hand equipment can remain with a dealer for a number of months, or even years, so there is normally plenty of time for due diligence. Some of the main issues are:

- Manuals/documentation: Make sure that you have these in place before buying second-hand equipment.
- Spares: It is important to establish that spare parts will be available for the machine.
- Condition/refurbishment cost.
- Control system: Many original equipment manufacturers put a great deal of time into the development of their control systems and it may be difficult to replicate.

The original suppliers of equipment offer documentation, training, mechanical warranties and, depending upon negotiations, performance guarantees. However, there is a clear tension between their options – creating a bargain for you is not as good as selling you a new unit, so it may not be routine to extract the same if you buy a second-hand unit.

13.2.4 Slurry and Application Questionnaire

Most filter original equipment manufacturers will begin their sales process by sending a questionnaire. It would be a good idea to have one prepared in advance of this, so that you highlight the issues that are important to you. In this way, you will be ahead of the game. It is important to be open and to include as much information as you can at this stage.

There will be a distinction between companies that offer a number of filter types and those that only offer one or two. There will inevitably be a tendency for filtration houses to offer the product that makes the most sense for them (in terms of profit and spares income), so it makes sense to check all of the possibilities.

A blank version of the form in Figure 13.1 is given in Appendix C (Figure C.1).[5]

13.2.5 After Commissioning

As discussed in Chapter 10, each filtration process is unique and there will be particular issues with your process that may not have been anticipated by a mechanical engineer in a drawing office; for example, your particular filter cake may accumulate in a particular spot and drag against a moving cloth.

After a certain period of operation, a small project (or series of projects) to squash any bugs or snags in the filtration plant – investigating the benefits versus costs of minor equipment modifications – should be conducted. This may require the addition of, say, deflector plates to prevent cake from building up at a certain location, or a permanent spray bar to wash the cake away every so often.

13.3 PROCESS OPTIMIZATION

In these last chapters of the book, the aim has been to draw together the ideas presented so far, in order to suggest a project methodology for the optimization of a process. The aim is that this method provides a convenient way to organize your thoughts and actions, in order to maximize your outcomes and minimize the costs of realizing these outcomes.

You may choose to follow this route rigorously, but even using the three images shown in Figure 13.2 as an *aide-mémoire* should provide focus when needed.

The motivation for process optimization, ultimately, should be based upon a desire to reduce production costs, improve product quality, increase productivity

---

[5] This form is also available for download from www.solid-liquid-filtration.com

Vendor questionnaire

Date:

Contact details

Name	
Company	
Address	
Website	Try to include any links to the overall process, or other sites giving background
E-mail, telephone	

Application description

Industry	Food, pharmaceuticals, mining, metallurgy, chemical....
Description of the application	Highlight the overall process, particularly those steps immediately before and after the filtration step. Include a flow sheet, if available.
Throughput	In whatever terms you choose (tonnes solids per hour, m³ slurry/hour), annual capacity. Just be clear.

Slurry Description

Solids component

Particle size data	e.g. 80% less than 30 μm, ideally, attach a particle size distribution
Composition	Chemical formula, common name (e.g. kaolin)
Physical description	Crystalline, amorphous, brittle, soft, sharp. Ideally include micrographs.
Abrasiveness	If notably abrasive, highlight it here

Liquid component

Composition	Common name (water, ethanol...) or full chemical composition
Viscosity	If you have lab data, include it, together with temperature. Otherwise, if notably viscous, highlight it here

Slurry composition

Percentage solids (w/w)	If applicable, give a range
Temperature	If applicable, give a range

Objectives

Cake moisture	Give a realistic target. If you have an existing figure, include it.
Washing results	In whatever terms you choose (conductivity of re-slurried cake, ash after burning...)
Wash liquor consumpt.	In whatever terms you choose (m³ per tonne solids or per tonne of slurry). If you have an existing figure, include it.

Any other Comments

General	What stage are you at? Process Design, Project Execution, Process Improvement. Previous experiences. This is the place to be expansive.
Regulatory	FDA, ATEX, Ex-Proof, Volatile Organics...

FIGURE 13.1 Example of questionnaire.

or improve SHE performance (or any combination). It can take many forms, from troubleshooting (fixing a failing process) through to shaving a few cents from your production costs or increasing production by 2%.

Map out your filtration process outcomes and relate these to your overall process success.

1. Map out the outcomes of your filtration process(es).
2. Relate these outcomes to the success of your overall process, perhaps using the Outcomes.xls sheet downloaded from www.solid-liquid-filtration.com.

Chapter | 13 Getting the Most from Filtration Processes

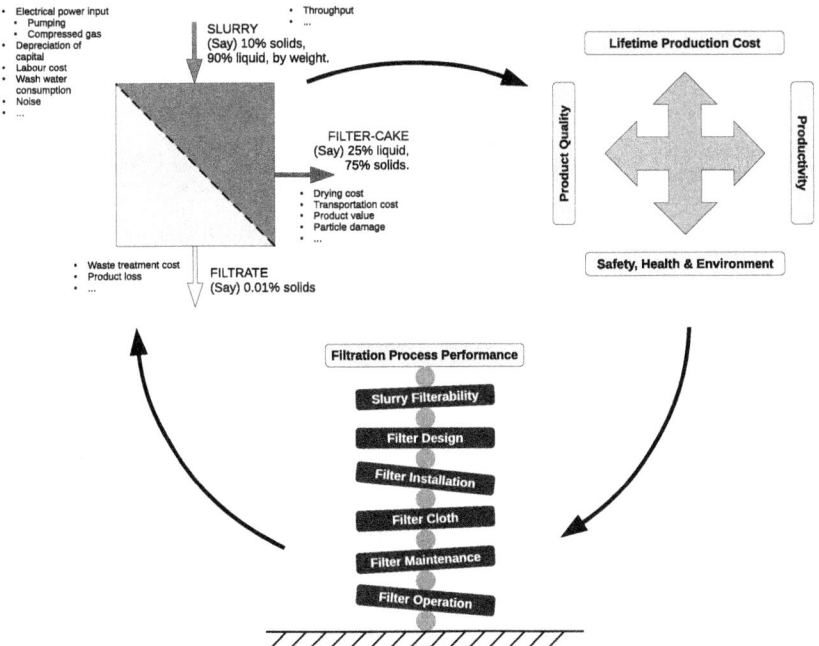

FIGURE 13.2 The central ideas of the book.

This may include drying costs, raw material costs and environmental performance.
3. Explore the benefits to be gained: conduct a sensitivity analysis; relate improvements against the anticipated cost to realize these improvements.
4. Explore each of the success factors for ways to achieve these improved filtration outcomes.

13.3.1 Optimization Projects

For anything more than a few minor tweaks to a process, a formalized project, with a team (two or more people) will often give a better result.

Ideally, if you are going to formalize a process optimization as a project, then a project with a clear name and objective will provide focus and motivation:

- "Reduce fuel costs by $1 per tonne to save $200,000 per year" is better than "reduce cake moisture by 5%".
- "Increasing productivity by 100 tonnes per year" is better than "find a filter cloth that gives better filtrate clarity".

The impact of filtration process outcomes on the overall process (Figure 13.3) should be investigated. In many cases, these impacts can be given a value (e.g. using the Outcomes.xls sheet at www.solid-liquid-filtration.com).

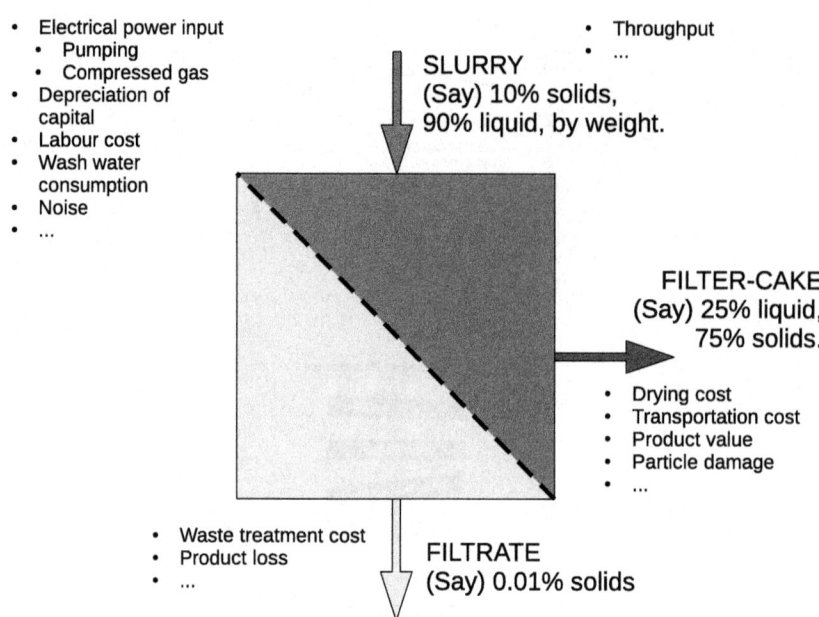

FIGURE 13.3 A more detailed description of the outcome of a filtration process.

Projects tend to succeed when they have a clear plan, even if it is on one page, giving:

- a clear set of goals, in terms that have real meaning (in money terms) rather than a more abstract meaning
- a clear team
- clear actions
- a clear end date.

13.3.1.1 Assembling a Team: Appointing a Champion

A visible process optimization project team leader is important. This person has the job of overcoming the skeptics and communicating the benefits of the team's work clearly and to report on progress. Equally, it is useful to have other interested parties involved, so that maintenance engineers and operators have an input into any suggested changes.

13.3.2 The Tower of Filtration Process Success Factors

The performance of a filtration process (viewed from the perspective of the overall process success) is determined by the combination of factors shown in Figure 13.4. Each of these was discussed at length in Part III.

Chapter | 13 Getting the Most from Filtration Processes   179

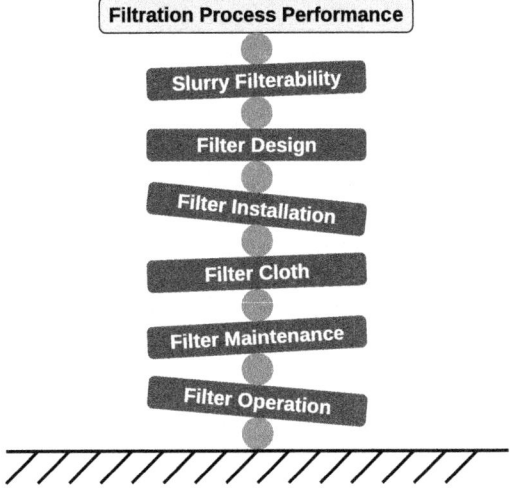

FIGURE 13.4 Tower of filtration process success factors.

The best way to optimize the process is to look at each of these factors. In general, the time taken to make changes to each factor increases with its position in the tower. It can take a matter of days (and good agreement-making skills) to make changes to the operation of a filter plant, weeks to change cloth, months to make meaningful changes to an installation and, potentially, years to replace a filter.

13.4 SUMMARY

It is to be hoped that by understanding how the outcomes of a filtration process affect the success of the overall process and, then, what factors determine these outcomes, you will be able to improve the success of the filtration process and, finally, the competitiveness of the overall process. These improvements may be far from obvious and may rely upon careful consideration of *what could be happening* inside the filter device, followed by testing of ideas on a small scale.

While these improvements may involve anything from a small change in operational procedures to an entire rethink of the filtration process (including the purchase of a different filter or type of filter), they can make a very significant difference to the success of your overall production process.

And I hope that this book helps.

Bibliography

Borges, A., & Aldi, J. (2009). Using a statistical model in the red mud filtration to predict the caustic concentration in the red mud. *Light Metals*, 2009, 117–120. Minerals, Metals & Materials Society.

Häkkinen, A., Pöllänen, K., Reinikainen, S -P., Louhi-Kultanen, M., & Nyström, L. (2008). Prediction of filtration characteristics by multivariate data analysis. *Filtration*, 8(2), 144–153.

Hippocrates (400 BC). *On airs, waters and places* (translated by Francis Adams, 1926).

Lu, L. (2006). Simulation of the cake formation and growth in cake filtration. *Minerals Engineering*, 19(10), 1084–1097.

Nicolaou, I. (2003). Novel software helps to solve filtration problems. *Filtration and Separation*, 40(8), 28–33.

Rushton, A., Ward, A., Holdich, R., Rushton, A., Ward, A., & Holdich, R. (2000). *Solid–liquid filtration and separation technology*. Wiley-Vch.

Salmela, N., & Oja, M. (2006). Monitoring and modelling of starch cake washing. *International Journal of Food Science and Technology*, 41(6), 688–697.

Sparks, T. (2010). Alumina: filtration in the alumina production process. *Filtration and Separation*, 20–93.

Stickland, A., White, L., & Scales, P. (2011). Models of rotary vacuum drum and disc filters for flocculated suspensions. *AIChE Journal*, 57(4), 951–961.

Wakeman, R., & Tarleton, E. (1999). *Filtration: equipment selection, modelling and process simulation*. Elsevier.

Wakeman, R., & Tarleton, E. (2005a). *Solid/liquid separation: principles of industrial filtration*. Elsevier.

Wakeman, R., & Tarleton, E. (2005b). *Solid/liquid separation: scale-up of industrial equipment*. Elsevier.

Appendix A

Useful Expressions

It is very useful to use mass and volume balances when performing a number of calculations concerning solid–liquid filtration. It is useful to be comfortable in using these expressions when simulating processes or producing information for reports or online data presentation.

Some of these expressions are a bit too clunky to be typed into a calculator every time you need them, so I would recommend using either a programmable calculator or a spreadsheet (many of the expressions in this section can be downloaded from www.solid-liquid-filtration.com).

A.1 SLURRY RELATIONSHIPS

First of all, look at a quantity of slurry (Figure A.1), where ρ_0 is density, V is volume, φ is the volume ratio (volume of solids: overall volume), M is mass and s is the solids ratio, with subscript s, l and sl designating solid, liquid and slurry.

The densities of the slurry and liquid, ρ_{sl} and ρ_l, can be found quite easily using a measuring cylinder and laboratory scales, as can s, the mass fraction of solids in the slurry, by drying a sample of slurry in an oven.

From these, using the volume balance given in Figure A.1:

$$V_{sl} = V_s + V_l \tag{A.1}$$

and seeing that s is defined as:

$$s = \frac{M_s}{M_{sl}} \tag{A.2}$$

and

$$V = \frac{M}{\rho} \tag{A.3}$$

then the volume of solids in the slurry is:

$$V_s = \frac{sM_{sl}}{\rho_s} \tag{A.4}$$

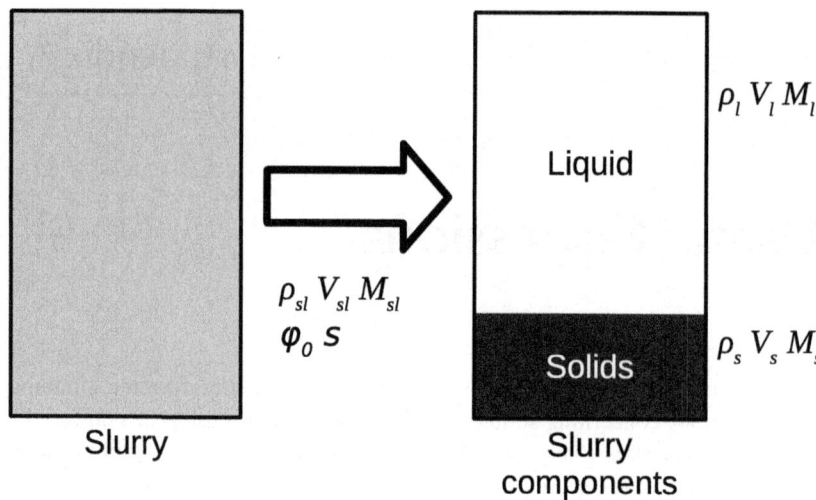

FIGURE A.1 Slurry broken down into its solid and liquid components.

and the volume of liquid is:

$$V_l = \frac{(1-s)M_{sl}}{\rho_l} \tag{A.5}$$

The volume balance (Equation A.1) can then be rewritten as:

$$\frac{M_{sl}}{\rho_{sl}} = \frac{sM_{sl}}{\rho_s} + \frac{(1-s)M_s}{\rho_l} \tag{A.6}$$

Cancelling out M_{sl} and rearranging, to get ρ_s alone on the left-hand side, we get:

$$\rho_s = \frac{s}{\left(\dfrac{1-s}{\rho_1} - \dfrac{1}{\rho_{sl}}\right)} \tag{A.7}$$

or, more simply:

$$\rho_s = \frac{\rho_l \rho_{sl} s}{\rho_{sl}(1-s) - \rho_l} \tag{A.8}$$

Now, in a very similar way, taking a mass balance from Figure A.1:

$$M_{sl} = M_s + M_l \tag{A.9}$$

the volume fraction of solids in the slurry is defined as:

$$\varphi_0 = \frac{V_s}{V_{sl}} \tag{A.10}$$

Appendix A

and seeing as:

$$M = V\rho \tag{A.11}$$

then the mass balance (Equation A.9) is:

$$(V_{sl})(\rho_{sl}) = \varphi_0(V_{sl})(\rho_s) + (1 - \varphi_0)(V_{sl})(\rho_l) \tag{A.12}$$

This simplifies, cancelling out V_{sl}, to give the volume fraction of solids in the slurry, in terms of the densities of slurry, solids and liquid:

$$\varphi_0 = \frac{(\rho_{sl} - \rho_l)}{(\rho_s - \rho_l)} \tag{A.13}$$

A.2 CAKE RELATIONSHIPS

Similar expression for the volume fraction in the cake can be established.

$$\varphi_c = \frac{(\rho_c - \rho_l)}{(\rho_s - \rho_l)} = 1 - \varepsilon$$

Where the subscript c denotes cake and ε is the porosity of the cake
See Figure A.2.

A.3 BALANCES ACROSS THE FILTER

The filtration process is summarized in Figure A.3.

To start with, the volume balance across the filtration process, assuming no solids in the filtrate, is:

$$\text{Volume slurry} = \text{Volume cake} + \text{Volume filtrate} \tag{A.14}$$

or

$$V_{sl} = LA + V \tag{A.15}$$

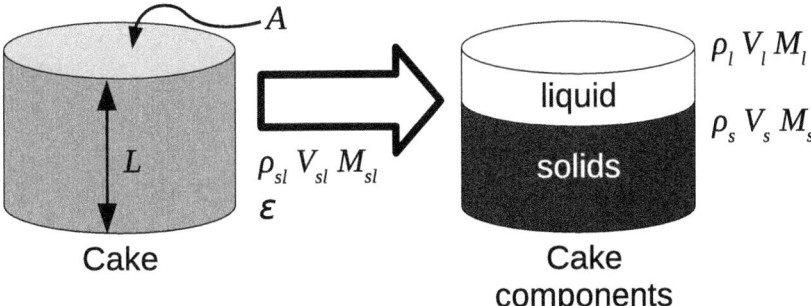

FIGURE A.2 Cake broken down into its solid and liquid components.

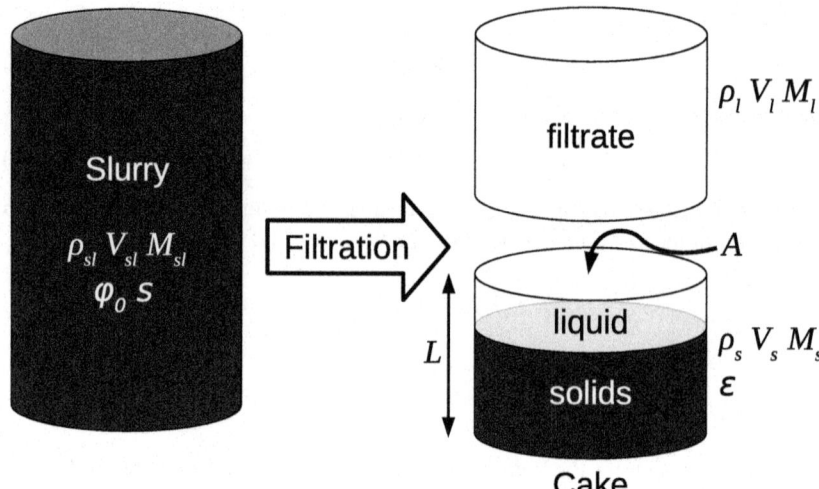

FIGURE A.3 The filtration process.

Also, the volume balance for liquid across the filtration process is:

Volume liquid in slurry = Volume liquid in cake + Volume liquid in filtrate (A.16)

First, the volume of liquid in the slurry is simply:

$$\underbrace{V_{liquid}}_{\text{in slurry}} = (1-\varphi_0)V_{sl} \qquad (A.17)$$

or (using A.15):

$$\underbrace{V_{liquid}}_{\text{in slurry}} = (1-\varphi_0)(LA+V) \qquad (A.18)$$

Then, the volume of liquid in the cake is:

$$\underbrace{V_{liquid}}_{\text{in cake}} = \varepsilon LA \qquad (A.19)$$

So, the volume balance (Equation A.16) becomes:

$$(1-\varphi_0)(LA+V) = V + \varepsilon LA \qquad (A.20)$$

Simplifying this, we get to:

$$L = \frac{V\varphi_0}{A(1-\varepsilon-\varphi_0)} \qquad (A.21)$$

Appendix A

Later on, it is very useful to be able to express the mass of cake solids per unit volume of filtrate, c:

$$c = \frac{\text{Mass cake solids}}{V} \quad (A.22)$$

so,

$$\text{Mass cake solids} = cV \quad (A.23)$$

but the mass of cake solids, from before is:

$$\text{Mass cake solids} = LA(1-\varepsilon)\rho_s$$
$$= \frac{V\varphi_0}{A(1-\varepsilon-\varphi_0)} A(1-\varepsilon)\rho_s \quad (A.24)$$

So, combining Equations A.23 and A.24 and simplifying:

$$c = \frac{\varphi_0 (1-\varepsilon)\rho_s}{(1-\varepsilon-\varphi_0)} \quad (A.25)$$

or, since $\varphi_c = 1 - \varepsilon$:

$$c = \frac{\varphi_0 \varphi_c \rho_s}{(\varphi_c - \varphi_0)} \quad (A.26)$$

A.4 SUMMARY

The following expressions can be useful in the laboratory or when figuring out production rates. First, the density of solids in a slurry (beware of gas in the slurry leading to an incorrect answer):

$$\rho_s = \frac{\rho_1 \rho_{sl} s}{\rho_{sl}(1-s) - \rho_1}$$

The volume ratio of solids in a slurry (useful if you already know the density of the solids from earlier):

$$\varphi_0 = \frac{(\rho_{sl} - \rho_1)}{(\rho_s - \rho_1)}$$

And the mass of solids per unit volume of filtrate collected from a filter:

$$c = \frac{\varphi_0 \varphi_c \rho_s}{(\varphi_c - \varphi_0)}$$

Appendix B

Flow Through a Growing Porous Filter Cake

Chapter 3 presented the idea that the flow of liquid through a filter cake becomes more difficult as the cake thickness increases, either requiring a greater pressure force to maintain a constant flow rate or, at a constant pressure, giving a reduced throughput during a filtration cycle.

The purpose of this appendix is to provide supporting mathematical background for those who are curious and to provide equations that can be used to model the cake formation stage in pressure or vacuum filtration processes.

B.1 FILTRATION EQUATION

The starting point comes from Darcy's law, established by Henry Darcy in the 1850s, following experiments passing liquid through a column of sand. This states that the flow rate through a porous bed is proportional to the differential pressure across the bed, the area of the bed and its permeability, and inversely proportional to the thickness of the bed and the viscosity of the fluid:

$$Q = \frac{kA \, \Delta p}{\mu L} \quad \text{(B.1)}$$

where Q is flow rate ($m^3 s^{-1}$), k is a permeability constant (m^2), A is area (of the filter) (m^2), Δp is differential pressure ($m^{-1} kg\, s^{-1}$), μ is viscosity ($m^{-1} kg$), and L is cake thickness (m).

Rearranging Equation B.1 to give pressure drop (Δp) in terms of the other quantities, we get:

$$\Delta p = \frac{Q\mu}{A} \left(\frac{L}{k}\right) \quad \text{(B.2)}$$

This equation applies if the flow through the porous bed (in our case, through the filter cake) is laminar, as it nearly always is in industrial applications. The L/k term is a resistance factor, and it clearly grows if L, the thickness of the bed, increases.

The pressure difference occurs over two distinct zones in the filter: through the filter cake itself and also through the thin zone where the filter cake and the filter cloth interact. We can write the total change in pressure, Δp, across these two zones as:

$$\Delta p = \Delta p_{\text{cake}} + \Delta p_m \tag{B.3}$$

It is important to note that the pressure drop across the filter medium, Δp_m, occurs when the fibers and pores in the filter cloth have been coated, and sometimes penetrated, by particles from the slurry. It does not refer to the resistance that would be offered to clear mother liquor by a clean cloth.

The following sections will look at each of these pressure drop terms in sequence.

B.1.1 Pressure Drop Over Cake: Δp_{cake}

$$\Delta p_{\text{cake}} = \frac{Q\mu}{A}\left(\frac{L}{k}\right)_{\text{cake}} \tag{B.4}$$

Now, the cake resistance term, (L_{cake}/k), as noted above, grows as the cake thickness increases. A volume balance was used across the filtration process to find L in terms of the volume of filtrate, V, and the slurry solids volume ratio, $\varphi_{(0)}$ and cake porosity ε in Equation A.21:

$$L = \frac{V\varphi_0}{A(1-\varepsilon-\varphi_0)}$$

In other words, providing the solids density in the slurry remains constant (which would be incorrect if the solids settle rapidly compared to the filtration time) and the cake porosity remains constant, then the cake thickness is proportional to the amount of filtrate collected.

So, we could substitute this expression for L into Equation B.4 and slightly later, in effect, we will. However, it is conventional for filtration technologists to use a slightly modified form for the resistance. Tabulated values for cake resistance given in reference books are usually given as a specific cake resistance, α, defined so that:

$$\left(\frac{L}{k}\right)_{\text{cake}} = R_{\text{cake}} = \frac{\alpha M}{A} \tag{B.5}$$

where M is the mass of solids in the filter cake, which we also looked at in Appendix A (Equation A.23):

$$\text{Mass cake solids} = cV$$

where c is the mass of solids in the cake per unit volume of filtrate (and is given by Equation A.25).[1]

---

[1] $c = \dfrac{\varphi_0(1-\varepsilon)\rho_s}{(1-\varepsilon-\varphi_0)}$

Appendix B

So:

$$R_{cake} = \frac{ac}{A}V \qquad (B.6)$$

and then:

$$\Delta p_{cake} = \frac{Q\mu ac}{A^2}V \qquad (B.7)$$

B.1.2 Pressure Drop Over Medium: Δp_m

Treating the filter cloth/first tiny layer of cake region as a thin bed that, under laminar flow conditions, can be described using Darcy's law, gives:

$$\Delta p_m = \frac{Q\mu}{A}\left(\frac{L}{k}\right)_m \qquad (B.8)$$

We can usually assume that the term $(L/k)_m$ is constant during the filtration cycle and, therefore, we can write it as R_m, so:

$$\Delta p_m = \frac{Q\mu}{A}R_m \qquad (B.9)$$

B.1.3 Pressure Drop Over the Cake and Medium

Substituting Equations B.7 and B.9 into Equation B.3:

$$\Delta p = \frac{Q\mu ac}{A^2}V + \frac{Q\mu}{A}R_m \qquad (B.10)$$

Now, since, a, c, μ, A and R_{cloth} are constants for a particular filtration cycle, we can rewrite this (just to make it quicker to write down) as:

$$\Delta p = Q(aV+b) \qquad (B.11)$$

where a represents the cake resistance factor:

$$a = \frac{ac\mu}{A^2}$$

and b the cloth resistance:

$$b = \frac{\mu}{A}R_{cloth}$$

So, rearranging:

$$Q = \frac{\Delta p}{(aV+b)} \qquad (B.12)$$

B.2 FILTRATION UNDER CONSTANT PRESSURE

If we operate a filtration cycle under constant pressure, then it is clear, from Equation B.12, that the flow rate through the filter (cake and cloth) will start relatively high and then fall as the amount of filtrate collected, and hence the amount of solids deposited in the cake, increase (Figure B.1).

If we zoom in on a part of this curve (Figure B.2), we can see that, after an interval of time, δt, we will have collected an amount of filtrate $\delta v \, (= Q \delta t)$, which we add to V, so that Q reduces, and so on. In fact, if we write Q in its

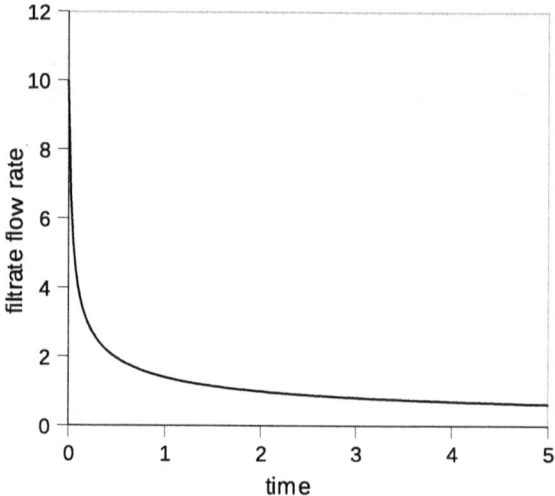

FIGURE B.1 Flow versus time through a growing filter cake.

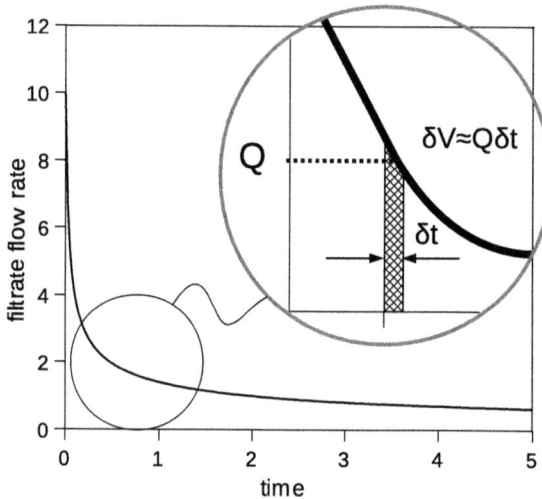

FIGURE B.2 Area under the curve.

Appendix B

differential form (dV/dt), by rearranging and applying boundary conditions we get:

$$\int_0^t dt = \frac{1}{\Delta p} \int_0^V (aV + b) dV \qquad (B.13)$$

so:

$$t = \frac{aV^2}{2\Delta p} + \frac{bV}{\Delta p} \qquad (B.14)$$

This gives the time required, t, to collect a volume of filtrate, V.

Replacing a' and b with the collections of constants that make them up:

$$t = \frac{\alpha c \mu}{2A^2 \Delta p} V^2 + \frac{R_{cloth} \mu}{A \Delta p} V \qquad (B.15)$$

Now, it is common to assume that the cloth resistance factor bV is very small compared to the cake resistance terms, especially when a large volume of filtrate has been collected (and V^2 is large). If this is the case:

$$V \approx \sqrt{\frac{2t}{a}} \qquad (B.16)$$

To produce double the amount of filtrate (or double the cake thickness) you need to multiply the time by 4. This can be checked easily in the laboratory and can be used at the full scale when considering how to operate your filters (times, speeds, etc.). This is shown in Figure B.3 (in this chart, Area 1 ≈ Area 2).

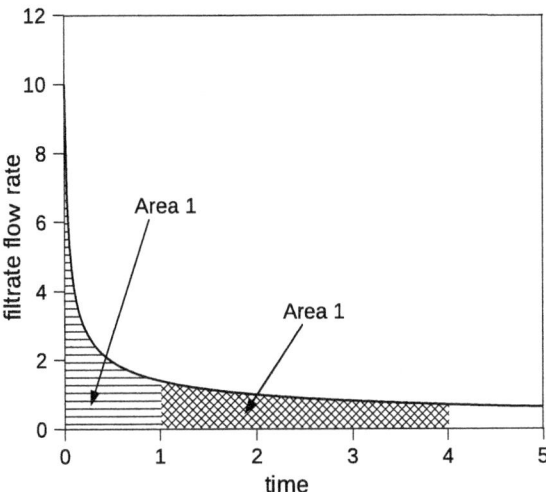

FIGURE B.3 Volume of filtrate collected versus time through a growing filter cake, at constant pressure.

B.3 FILTRATION WITH CONSTANT FLOW

Under other circumstances, a filter may be operated under conditions of constant flow. This can be achieved using either a positive displacement pump running at a constant speed (being careful to avoid pulsating flow and pressures), or a variable-speed drive pump and control system.

Going back to Equation B.10:

$$\Delta p = Q \left(\frac{\mu \alpha c}{A^2} V + \frac{\mu}{A} R_m \right)$$

If the filter cake is incompressible, then differential pressure above and below the cake Δp will simply increase in proportion with the volume of filtrate collected, V (the only variable on the right-hand side of the equation).

Appendix C

Forms and Templates

The forms and templates given in this appendix can be downloaded from www.solid-liquid-filtration.com

Vendor questionnaire   Date:

Contact details
- Name
- Company
- Address
- Website
- E-mail, telephone

Application description
- Industry
- Description of the application
- Throughput

Slurry Description

Solids component
- Particle size data
- Composition
- Physical description
- Abrasiveness

Liquid component
- Composition
- Viscosity

Slurry composition
- Percentage solids
- Temperature

Objectives
- Cake moisture
- Washing results
- Wash liquor consumpt.

Any other Comments
- General
- Regulatory

FIGURE C.1 Example of questionnaire.

195

Reference Visit Checklist   Date:

Location

Company	
Address	
Website	
E-mail, telephone	
Persons present	

1.	Are you speaking to the original decision maker? If so, how was the decision justified?	
2.	How long has the filtration plant been operating?	
3.	Would the user recommend this type of filter?	
4.	What problems were experienced?	
5.	How is spare part delivery?	
6.	What other alternatives were considered at the time of purchase?	
7.		
8.		
9.		
10.		
11.		
12.		
13.		
14.		
15.		
16.		
17.		
18.		
19.		
20.		

FIGURE C.2 Example of reference visit checklist.

Appendix C

Assessment of Change   Date:

Proposed Change

Result of change

Cost	Quality

SHE	Productivity

FIGURE C.3 Example of change assessment.

Appendix D

Sample Test Method

The method below is taken from a real test report. It is not intended to be a recipe, but should give a reasonably suitable layout and form.

ASSESSMENT OF SLURRY FILTERABILITY

Objective

A simple, repeatable method for the assessment of (slurry) filterability. This is to allow the root cause of filtration performance to be determined; for example, if filtration rate is poor, is it slurry filterability or other plant issues (vacuum, scaling, cloth condition, take-off roller/combs) that are responsible?

Background

This test method deliberately uses a filtration time that is greater than that on the filters in operation on the plant, because it is very difficult to reproduce the short filtration times from the plant, reliably, in the laboratory. Even though this method will produce filter cakes approximately 10–12 mm thick (compared with approximately 2 mm on the full-scale filter) it is important to note that each 2 mm layer in the thicker cakes will have the same filtration properties as the cake on the plant filters, so this is a valid method.

The method also assumes that the cake solids volume fraction remains the same for all muds tested. This can be confirmed by knowing the solids content of the slurry and the amount of slurry added, and measuring the cake thickness (when the upper surface of the cake just begins to dry, so that no air displacement of liquor has occurred). From these data, it is possible to determine the cake solids volume fraction.

Apparatus

The following apparatus is to be used:
- Büchner funnel
- laboratory scales, measuring to 0.01 g
- graduated vacuum cylinder
- stopwatch

- permanent marker pen
- 1 liter bottle for each test
- clock-glass for each test (able to hold 100 g of slurry)
- water bath, set to 30°C
- vacuum source (ejector)
- filter cloth (use the actual filter cloth from the full-scale filter)
- filterability.xls spreadsheet (a data sheet for inputting test results and producing a report on the mud filterability; Figure D.1).

Set-up

Sample collection and preparation

1. Collect the sample from the underflow of thickener in 1 liter sample bottles and transport it to the laboratory as soon as possible.
2. Divide the slurry into 450 g sample batches, ensuring that all samples are approximately the same (±20 g).
3. Put the samples into the bottle roller, set to 30°C and leave long enough for the temperatures to stabilize.
4. Using the remaining sample, measure the slurry density, weighing approximately 100 ml of slurry in a graduated measuring cylinder and recording the volume and weight.

Method

Preparation for a test

1. Apply a small quantity of silicone grease to the ground-glass surface on the top of the graduated cylinder, to ensure that there is a good seal against the vacuum.
2. Assemble the Büchner filter, by placing the filter cloth centrally onto the support grid and locating the filter collar in place using the four clamps.
3. Fix a short piece of silicone tubing to the bottom of the Büchner funnel to direct the flow of filtrate and to reduce splashing and bubble formation in the cylinder.
4. Place the Büchner funnel on top of the vacuum flask and check that it is level, with the tubing in the graduated cylinder pointing away from the graduations in the cylinder (so that the filtrate will not be disturbed as you take readings).
5. Turn on the air supply to the vacuum ejector.
6. Check the vacuum pressure (using the rubber bladder to stop the air flow through the filter cloth). Compare this to the pressure when the vacuum line is blocked (by pinching the tubing). The vacuum pressure should be set to 0.5 bar.
7. If there is a difference between the vacuum pressure with the rubber bladder blocking the cloth and the pressure when the tubing is closed then there is a leak of air in the set-up. Check for leaks and, if necessary, apply a little more silicone grease.
8. Once you are sure that there are no leaks, turn off the air supply to the vacuum ejector.

Appendix D

Test Data   Date: Performed by:

Project		
Objective		
Equipment		
Area		

			\multicolumn{9}{c}{Test number}
			1	2	3	4	5	6	7	8	9
Cloth											
Slurry	% solids	wt/wt									
	temperature	°C									
	pH										
	Density	gpl									
	Comment										
Filtration	time	s									
	pressure	bar									
	volume	ml									
	filtrate solids	ppm									
Washing	liquid name	-									
	time	s									
	pressure	bar									
	volume	ml									
	filtrate solids	ppm									
Pressing	time	s									
	pressure	bar									
	volume	ml									
	filtrate solids	ppm									
Air drying	time	s									
	pressure	bar									
	Flow-rate	lpm									
	volume	ml									
	filtrate solids	ppm									
Cake	weight	kg									
	thickness	mm									
	moisture	%w/w									
Filtrate	Volume	ml									
	filtrate solids	ppm									
Comments											
	Total time	s									
	Cake solids	kg									
	Capacity	kg/(m²h)									

General comments

FIGURE D.1 The filterability.xls spreadsheet.

Performance of a test

Note that the following procedure should be followed without interruption, to ensure consistent conditions for each test.

1. Remove a sample bottle from the roller bath.
2. Take a clock glass, label it with a test number and place on the weighing scales. Record the weight in the "data" tab on the filterability.xls spreadsheet.

3. Pour approximately 100 g of the slurry into the clock-glass. Record the weight in the next cell down in the same spreadsheet.
4. Place the clock-glass into drying oven, at 100°C. (After a period of several hours, the dried weight will be added to calculate the solid content of the slurry.)
5. Place the slurry bottle on the weighing scales, with the top screwed shut, and record the amount in the filterability.xls spreadsheet.
6. Making sure that the top is secure, agitate the contents of the slurry bottle to ensure that the contents are well mixed.
7. Pour the contents of the slurry bottle into the top of the Büchner funnel, ensuring that the slurry is distributed evenly (a gentle tap on the side of the filter may be necessary to make sure).
8. Open the air supply to the vacuum ejector and start the stopwatch when the vacuum pressure reaches the set level.
9. Record the time at which the lower meniscus of the filtrate reaches the top of the first division (the lowest) in the graduated cylinder in the filterability.xls spreadsheet. (NB. Because the markers in the cylinder are not perfectly level, use the left-hand side of the division.)
10. Record the time at which the lower meniscus of the filtrate reaches the top of the next three divisions in the graduated cylinder.
11. When you have recorded the time for the fourth division, the test is complete. (NB. You must fill in all four points for the spreadsheet to work.) You may choose to leave the vacuum on to dry the cakes and make it easier to wash up after each test. Alternatively, you may choose to run the test until the upper surface of the cake just begins to dry, to measure the cake thickness and determine the cake solids volume fraction.
12. Once the vacuum has been stopped, remove the Büchner funnel from the graduated cylinder and wash up the complete kit.
    a. Wash the filter cloth thoroughly and shake dry.
    b. Disassemble, wash and dry the filter parts.
    c. Rinse and dry the graduated cylinder.
13. Repeat the test procedure for each sample.

Analysis and interpretation of results

Once the data have been entered into the filterability.xls spreadsheet, a graph will be produced on the "results" tab. This graph shows how the filtration parameter compares with the standard mud conditions (given on the "master curve" tab). A mud sample that is harder to filter will produce a point above the master curve; conversely, a mud sample that is easier to filter than the standard conditions will produce a point below the curve.

Index

Page numbers followed by *f,* indicate figure and *t,* indicate table.

A
Acid rain, 17
Additives, 78–79
Air drying, 39–41, 40f
   hot gas drying, 41
Alumina, 67–71
   flowsheet, 68f
   hydrate filtration and washing, 71
   precipitation seed filtration, 70–71
   residue separation, 69–70
      liquor polishing, 70
      red mud filtration, 69–70
Aluminum, production growth of, 17, 18f

B
Batch centrifuges, 121–123, 122f
   applicability, 123
   cake discharge, 122
   cake washing, 122
   gas drying, 122
   installation, 123
   maintenance, 123
   options, alternatives and variations, 122–123
   slurry feeding/filtration, 121
Batch pressure filters, 157–158
Bauxite, 67
Bayer process, 67
Belt filtration, 17
   in pressurized housing, 103–104, 104f
Body feed, 78
   filtration, 43–44, 44f
Burr, in filter cloth, 142f
Buying and selling process, 172–174
   cloth suppliers, 173
   references, 173–174
   relationship management, 173

C
Cake. *See* Filter cake
Calendering, 138–139, 140f
Candle filters, 117–120
   air drying, 118–119
   applicability, 120
   cake discharge, 119
   cake washing, 118
   cloth cleaning, 119–120
   general arrangement of, 118f
   installation, 120
   options, alternatives and variations, 120
   slurry feeding, 118
Centrifugal filtration, 121–123
   batch centrifuges, 121–123, 122f
   continuous centrifuges, 123
Ceramec capillary action disc filters, 93–94
Chamber plates, 105, 108f–109f
Change assessment, 197
Chemical additives, 78
China clay, 14
Cloth. *See* Filter cloth
Coating, of filter cloth, 140, 140f
Competitiveness, in processing, 3, 4f, 49–54, 52f
   product quality, 51
   production cost, 50
   productivity, 51–52
   safety, health and environment, 52–53
Continuous centrifuges, 123
   applicability, 123
   cake discharge, 123
Control head, 86f
Conveying, filter cake, 60–61
Corn starch, modified, 34
Costs
   of drying, 57–58, 58f
   filtration, 63–64
      air and water, 63
      consumables, 64
      maintenance costs, 64
      operator costs, 64
      power costs, 63
   production, 50
   transportation, 59–60
Counter-current washing, 37–38, 37f

D
Darcy's law, 189
Data, test
   acquisition, 165–166
   archiving, 166, 167f

203

Dewatering, 64–67, 71–72
Diaphragm plates, 107, 108f
Diatomaceous earth, 44f
Drinking water, 13
Drying, cost of, 57–58, 58f
Dusting, filter cake, 61

E

Electroplating, 62–63
Electrostatic charge, surface, 29, 42
Energy crisis, 17
Environment. *See* Safety, health and environment (SHE)
Equipment selection, 170–171
    buying and selling process, 172–174
    after commissioning, 175
    principles of, 171–172
    second-hand equipment, 174–175
    slurry and application questionnaire, 175, 176f

F

Fabrication, of filter cloth, 141
Fermentation broth, 34
Filter, types of, 85f
Filter cake, 23
    air drying, 39–41, 40f
    compressible, 43
    draining, 38–39
    filtrate volume and filtration-step time, relationship between, 31, 32f
    fine particles, migration of, 43
    formation of, 26–29
    growth, 29–33, 29f
        and rate of flow, relationship between, 31, 31f
    handling, 129
    moisture, 57–61
    outcomes of, 57–61
    particle breakage, 61
    porous, flow through growing, 189–194
    pressing, 33
    properties, variation in, 41–42
        macroscopic and mesoscopic, 41–42, 42f
        microscopic, 41
    thickness, and pressure drop, 31–32
    washing, 33–38, 35f–38f, 61
Filter cloth, 23, 26–29, 26f–28f, 131–146
    cleaning, 144, 144f–145f
    design and manufacture, 133–141
        cloth finishing and fabrication, 141
        materials, 134–135, 134t
        postweaving treatment, 138–141
        weaves, 136–138
        yarns, 135–136
    desired outcomes, 133
    failure, 143–144
    repair, 144–146
    and slurry, interaction between, 27–28, 28f
    suppliers, 173
    support grid, 141–142, 141f
Filter design, 81–124
    centrifugal filtration, 121–123
    pressure filtration
        continuous, 103–104
        discontinuous, 104–121
    vacuum filtration, continuous, 84–103
Filter installation, 125–130
    cake handling, 129
    example for, 127f
    human considerations, 126–128
        access, 126
        flooring, 127–128
        lighting, 126–127
        noise, 126
        ventilation and extraction, 127
    process considerations, 128–129
    slurry systems, 129
Filter maintenance, 147–152
    component plant trials, 151
    issues with equipment, 148–149
    multiple-filter installation, optimal operating regime for, 149–151, 150t
    speed versus machine sympathy, 149–151
Filter operation, 153–158
    choices, 154–158
        batch pressure filters, 157–158
        effect of speed on top-fed filters, 157
        trough-fed filter capacity vs. speed, 154–157
Filter press, 15f–16f, 109
    air drying, 110
    applicability, 111
    cake discharge, 111
    cake pressing, 109
    cake washing, 109–110
    cloth cleaning, 111
    cycle, 110f
    elements of, 104–105
    overhead beam, 106f
    plate and frame pack, elements of, 107f
    side-bar, 106f–107f
    slurry feeding, 109, 110f

Index

Filtrate, 23
  clarity and volume of, 62–63
    precipitation/electroplating, 62–63
    product losses, 62
  concentration, and washing performance, 34–36, 36f
  outcomes of, 61–63
Filtration
  cycle, 23–24, 24f
    effect in dewatering application, 25f
  industry
    current trends of, 17–18
    recent history of, 16–17
    shape of, 18
  origins of, 13–14
  process
    basic narrative of, 23–24
    special cases in, 25–26
    step-by-step narrative of, 25–41
  types of, 21–22
Filtration and Separation Buyer's Guide, 172–173
Filtration equation, 189–191
  pressure drop over cake, 190–191
  pressure drop over cake and medium, 191
  pressure drop over medium, 191
Flow, liquid
  filtration with constant, 193
  speed of, 28, 42
Flue-gas desulfurization, 17

G

Gas, 23
Gravity forces, 29

H

Health. *See* Safety, health and environment (SHE)
Heat setting, of filter cloth, 141
Hippocrates, 13
Horizontal filter. *See* Pan filter
Hot gas drying, 41

I

Industrial applications, of filtration, 2
Industrial Revolution, developments in filtration industry after, 14–15
Inference, 22
Ishikawa diagram, of filtration performance success factors, 6f

K

Kaolin, 14
  particles, 25f–26f, 26–27

L

Laboratory scale testing equipment, 162, 163f
Landfill taxes, 60
Leaf filters, 121

M

Mathematical expressions
  for balances across filter, 184–185
  for slurry relationships, 183–184
Medium, filter, 26–29
Metal refinery residue, 34
Mineral concentrate, 64–67
Mining, 64–67
  crushing/grinding, 64, 65f
  filtration, 66
  flotation, 64–66, 66f
  process, 65f
Moisture, filter cake, 57–61
  binding/pelleting moisture, 60
  conveying, 60–61
  dryer throughput, 58–59, 59f
  drying, cost of, 57–58, 58f
  dusting, 61
  fuel consumption related to, 57f
  landfill taxes, 60
  transportation costs, 59–60
  transportation moisture limit, 60
Monofilament yarns, 135
Mother liquid, 23
Motive force, for filtration, 23
Multifilament yarns, 135–136
  pinhole in, 143f
Multilayer weaves, 138, 139f
Multiple-filter installation, optimal operating regime for, 149–151, 150t

N

Natural fibers, 134
Nylon. *See* Polyamide

O

Outcomes, of filtration processes, 3, 4f–5f, 6, 55–72, 56f, 178f
  examples, 64–72
  filter cake, 57–61
  filtrate, 61–63
  filtration costs, 63–64
  slurry, 63

P

Pan filter, 101–103
  air drying, 102
  applicability, 103
  cake discharge, 102
  cake washing, 101–102
  cloth cleaning, 102
  general arrangement of, 102f
  maintenance, 103
  options, alternatives and variations, 102
  slurry feeding/filtration, 101
Performance, of filtration, 6–7
  success factors, 6f–7f, 178–179, 179f
Physical testing, 161
Pilot-scale testing equipment, 163–164
Plain weave, 136–137, 136f
Plant trials, 151
Polyamide, 134
Polyester, 134
Polymers, 134
Polypropylene, 134
Polytetrafluoroethylene (PTFE), 134
Polyvinylidene fluoride (PVDF), 134
Potato starch particles, 25f, 26–27
  volume-based particle size distribution of, 30f
Precipitation, 62–63
Precoat filtration, 43–44, 44f
Prefiltrate, 29–30
Pressure filtration, 17
  under constant pressure, 192–193
  continuous, 103–104
  discontinuous, 104–121
    candle filters, 117–120
    filter press, 109
    leaf filters, 121
    spinning disc filters, 120–121
    tower press, 111–114
    tube press, 115–117
Prestretching, of filter cloth, 141
Process design, 170–175
Process optimization, 175–179
  projects, 177–178
    team, assembling, 178
Process testing, 161–168
  analysis, 166–168
    scaling up, 168
  cake washing, 166
  data acquisition, 165–166
  data archiving, 166, 167f
  data sheet, 167f
  design of experiments, 164–165

  example method, 165
  sampling, 165
  sample test method, 199–202
  test equipment, 162–164
    laboratory scale, 162, 163f
    pilot-scale, 163–164
  testing program, 164
Product development, 170
Production cost, 50
Productivity, 51–52
Product quality, 51

Q

Quality, product, 51
Questionnaire, sales, 175, 176f, 195

R

Raw materials, 50
Reciprocating tray filter, 94, 95f
Reference visits, 173–174
  checklist, 196
Reflux washing, 38, 38f
Relationship management, 173
Replacement washing, 34, 35f
Root-cause diagram. See Ishikawa diagram
Rotary pressure filter, 104, 105f
Rotary vacuum disc filter, 91–94
  air drying, 92
  applicability, 93
  cake discharge, 93
  cloth cleaning, 93
  general arrangement of, 92f
  installation, 93, 128f
  maintenance, 93
  options, alternatives and variations, 93–94
  washing, 92
Rotary vacuum drum filter (RVDF), 15f, 84–91, 85f, 87f–88f
  air drying, 88
  applicability, 90
  cake discharge, 88–89
  cake pressing, 87
  cake washing, 87
  cloth cleaning, 89
  configurations, 89f
  installation, 89
  maintenance, 90
  options and alternatives, 90
  precoat filters, 90–91
  slurry feeding/filtration, 86–87
Rubber belt vacuum filter, 99–101

Index

S

Safety, health and environment (SHE), 52–53
Sampling, 165
Satin weave, 138, 139f
Second-hand equipment, 174–175
Slurry, 22–23, 25–26
   and cloth, interaction between, 27–28, 28f
   filterability of, 75–80
      assessment of, 199–202, 200f
   handling, 79–80
      flow control valves, 80
      pumping, 79
      storage/suspension, 79–80
   nature of, 76
   outcomes of, 63
   pretreatment of, 76–79
      additives, 78–79
      density, 77
      temperature, 77–78
   pumping systems, 129
   rapidly settling, 42
Slurry filtration equipment, 2
Solid-liquid suppression. *See* Slurry
Solid particles, 25–29, 25f–28f
   rigidity of, 78
   shape of, 29
Spinning disc filters, 120–121, 121f
Starch washing, 71–72
Stop–go filter, 95
Support grid, for filter cloth, 141–142, 141f

T

Table filter. *See* Pan filter
Thought experiment, 22
Top-fed filters, effect of speed on, 157
Tower press, 111–114
   air drying, 112
   applicability, 114
   cake discharge, 114, 114f
   cake pressing, 112
   cake washing, 112
   cloth arrangements in, 112
   cloth cleaning, 114
   options, alternatives and variations, 114
   outcomes of, 112
   plate, 113f
   single-cloth, general arrangement of, 113f
   slurry feeding, 112
Transportation costs, 59–60
Transportation moisture limit (TML), 60
Tray type vacuum belt filter, 94–99

Trough-fed filter
   capacity vs. speed, 154–157
   multiple filter scenarios, 155–157, 156t
Tube press, 115–117, 115f
   air drying, 116
   applicability, 117
   cake discharge, 116–117
   cake washing, 116
   cloth cleaning, 117
   installation, 116f, 117
   maintenance, 117
   options, alternatives and variations, 117
   slurry feeding/filtration, 116
Twill weave, 137–138, 137f

V

Vacuum belt filter
   rubber belt type, 99–101
      air drying, 99
      applicability, 100
      cake discharge, 100
      cake washing, 99
      cloth cleaning, 100
      general arrangement of, 99f
      installation, 101
      maintenance, 100–101
      options, alternatives and variations, 100
      slurry feeding/filtration, 99
   tray type, 94–99, 94f, 98f
      air drying, 97
      applicability, 98
      cake discharge, 97, 97f
      cake pressing, 96–97
      cake washing, 96
      cloth cleaning, 97–98
      installation, 98
      maintenance, 99
      options, alternatives and variations, 98
      slurry feeding/filtration, 96
      types of, 94
      with multiple washing stages, 96f
Vacuum drum filters, multiple filter scenarios, 155–157, 156t
Vacuum filtration, continuous, 84–103
   pan filter, 101–103
   rotary vacuum disc filter, 91–94
   rotary vacuum drum filter, 84–91, 85f, 87f–88f
   vacuum belt filter
      rubber belt type, 99–101
      tray type, 94–99

Variable volume chamber filters, 157–158
Viscosity of liquid, 28, 42, 78

W
Washing, filter cake, 61
Wash liquid, 23
Waste, as raw material, using, 50
Water, drinking, 13
Weaves, 136–138

multilayer weaves, 138, 139f
plain weave, 136–137, 136f
satin weave, 138, 139f
twill weave, 137–138, 137f
Wheat starch particles, volume-based particle size distribution of, 30f

Y
Yarns, 135–136, 135f